Differential Manifolds

With a New Appendix by John W. Morgan
on the Work of Grigori Perelman

Antoni A. Kosinski

Department of Mathematics
Rutgers University
New Brunswick, New Jersey

Dover Publications, Inc.
Mineola, New York

Bibliographical Note

This Dover edition, first published in 2007, is an unabridged, slightly corrected republication of the edition originally published by Academic Press, Boston, in 1993. The present edition includes a new Appendix by John Morgan on the work of Grigory Perelman (see Appendix II, pages 233 to 246).

Library of Congress Cataloging-in-Publication Data

Kosinski, Antoni A., 1930–
    Differential manifolds / Antoni A. Kosinski.
        p. cm.
    "This Dover edition, first published in 2007, is an unabridged, slightly corrected republication of the work originally published by Academic Press, Boston, in 1993."
    ISBN-13: 978-0-486-46244-8
    ISBN-10: 0-486-46244-7
    1. Differential manifolds. I. Title.

QA614.3K66 2007
516.3'6—dc22

                                                                    2007018107

Manufactured in the United States of America
Dover Publications, Inc., 31 East 2nd Street, Mineola, N.Y. 11501

*This book is dedicated to the memory of my father,
the founder and publisher of "Mathesis,"
the first scientific publishing house in Poland.*

April 1963 Symposium in honor of Marston Morse,
Institute for Advanced Study, Princeton, New Jersey.

First row, seated, from left: S. S. Chern, R. J. Pohrer, A. Selberg,
M. Morse, W. Leighton, M. Hirsch, S. S. Cairns, H. Whitney.
Second row, standing, from left: R. Bott, B. Mazur, G. A. Hedlund, T. Frankel,
S. Smale, N. Kuiper, J. F. Adams, W. Browder, J. W. Milnor, M. Kervaire.

Contents

Introduction

Three decades ago differential topology went through a period of extremely rapid growth. Six years after Milnor's discovery of non-equivalent smooth structures on a topological sphere such structures were already classified through the newly invented method of surgery. In the same period Smale showed that every manifold can be constructed by successive attachment of handles and provided a method to obtain the most economical description. This enabled him to prove the Poincaré conjecture in higher dimensions, thereby demonstrating the strength of his methods. Also in the same period, Smale, Haefliger, and Hirsch developed the theory of imbeddings and immersions, vastly extending the foundational results of Whitney.

The methods invented in these early years subsequently gave rise to a large amount of research. Thus the handle constructions of Smale were extended by Kirby and Siebenmann to topological manifolds, and the method of surgery was expanded and applied successfully to a large variety of problems by Browder, Novikov, and Wall.

Looking back at these developments it appears possible at present to single out certain main ideas and results and present them in a systematic and consolidated way. This is what I have attempted to do here. In a very broad way, the content of this book can be described as the study of the topological structure of smooth manifolds. As I intended this book to be accessible at the beginning of graduate studies, this is preceded by a presentation of basic concepts and tools of differential topology.

An overview of the material follows. More precise information is in the introductions to the individual chapters.

The first two chapters introduce in moderate detail the notions of smooth manifold, submanifold, and tangent space. Imbeddings are discussed briefly and isotopies at length. The presentation is complete, but it is assumed, implicitly, that the subject is not totally unfamiliar to the reader. A more leisurely treatment of the analytical topics of these two chapters can be found in [Bo].

In Chapter III it is shown that a neighborhood of a submanifold of a smooth manifold can be fibered by planes, that is, it is a vector bundle, and that this bundle structure is unique. This is a fundamental result and the basis for all that follows.

The concept of transversality due to Thom is introduced in Chapter IV. This is the smooth counterpart of the notion of general position and is used similarly to extract from messy entanglements their essential geometric content. In this chapter it is applied to prove that every function can be approximated by one with a very regular behavior at singularities, a Morse function. It is also used to define intersection numbers. This geometric concept will in later chapters supplant the less intuitive cup product.

The results of Chapter V are not utilized elsewhere in this book. It provides an introduction to the beautiful and difficult theory of foliations.

These first four, or five, chapters constitute a general background not only for differential topology but also for the study of Lie groups and Riemannian manifolds. The analytical means employed here have their roots in the implicit function theorem, the theory of ordinary differential equations, and the Brown–Sard Theorem. Some algebraic results in the form adapted for the purpose and collected in the appendix are used as well. Very little algebraic topology enters the picture at this stage.

Chapter VI is devoted to a description of various ways of gluing manifolds together: connected sum, connected sum along the boundary, attachment of handles, etc. The presentation avoids the usual smoothing of corners. There is a brief discussion of the effect of these operations on homology; it prepares the ground for the more precise results of the following chapters. The last two sections describe a way to build some highly connected manifolds; it is shown in Chapter VIII that all highly connected manifolds can be constructed in this way.

An important result of Chapter VI describes the situation when two successive attachments of handles produce no change: The second handle destroys the first. This is Smale's Cancellation Lemma. Chapter VII begins with the proof that every manifold can be built by a successive attachment of handles of increasing dimension. To such a structure there is associated a chain complex yielding the homology of the manifold. The chains are linear combinations of handles and the boundary operator is given by a matrix of intersection numbers. Of course, the same is true for a triangulation or a cellular decomposition, but the relation between the handle presentation and the homology structure of the manifold is very transparent geometrically. This fact is exploited here to obtain a simple proof of the Poincaré duality theorem and of the Morse inequalities providing the lower limit for the number of handles necessary to construct a manifold with given homology. The 3-dimensional case, at the end of the chapter, provides a nice illustration of basic ideas.

Chapter VIII contains the proof of the existence of a handle presentation with the minimal number of handles determined by its homology groups. The following example should explain the importance of this idea. The minimal number of handles necessary to build an n-dimensional sphere is two: two n-discs glued along boundaries. If we succeed in proving that a homotopy sphere admits a presentation with the minimal number of handles determined by its homology, then it must admit a presentation with two handles. In turn, this implies that it is homeomorphic to the sphere, i.e., the Poincaré conjecture.

The proof we give here follows the original idea of Smale to manipulate handles, not the Morse functions, and adopts the following point of view. The homology of the manifold is given by chain groups (generated by handles) and homomorphisms described by matrices of intersection numbers. It is well-known how this structure can be reduced through a sequence of algebraic operations to the most economical form, for instance, with all matrices diagonal, etc. We try to find geometric operations on handles that are reflected by these algebraic operations on their algebraic counterparts: the generators of chain groups. The key to success is in the cancellation lemma, which, together with Whitney's method of eliminating unnecessary intersections, permits the actual geometric elimination of those handles whose presence is algebraically superfluous.

The use of Whitney's method necessitates dimensional restrictions. The final result asserts in its simplest form the existence of the minimal presentation for simply connected manifolds of dimension higher than 5. This has a large number of consequences: the h-cobordism theorem, the Poincaré conjecture, and the characterization of the n-disc and of highly connected manifolds being among the most important.

Chapter IX presents a construction invented by Pontriagin that associates to a framed submanifold of codimension d of a manifold M a map of M into a d-sphere, and to a suitably defined equivalence class of such submanifolds the class of homotopic maps. This provides a link between homotopy theory and differential topology. Various operations known from homotopy theory can be represented by geometric constructions; as an illustration we utilize this method to provide proofs of some classical theorems of Hopf and Freudenthal.

Next, we turn our attention to the class of manifolds that admit framings, that is, manifolds that can be imbedded in a Euclidean space with a trivial normal bundle. In the last section it is proved that homotopy spheres have this property. This result is crucial for the classification of differential structures on spheres in the next chapter. In the proof of it we must invoke, for the first time in this book, some deep results of Adams, Bott, and Hirzebruch.

Chapter IX can be read directly after Chapter IV.

The last chapter introduces the method of surgery. An overview of it is given in the introduction to the chapter and it is too involved to give here. The main line of argument follows the classical paper of Kervaire and Milnor, but is simplified through the use of the theory of handle presentation from Chapter VIII. This chapter closes with the classification of smooth structures on spheres in terms of stable homotopy groups of spheres and a few examples of nonstandard structures.

The theory of imbeddings is not considered here, though the classical results of Whitney are quoted and used. It stands somewhat apart from the subjects considered here and I did not want to expand unduly what was to be a short book requiring minimal prerequisites. The book is somewhat longer than intended but the prerequisites remain limited to what is usually found in a first course in algebraic topology, and to elements of the theory of vector bundles. The theory of cohomology products is not used until the last chapter, and even there it could be dispensed with.

Various shortcuts are possible in reading this book. The reader who wishes to proceed quickly to Smale's theory can skip the last two sections of Chapters III and VI, as well as Chapter V. The surgery method of the last chapter is accessible after elements of Chapters VI and IX, provided that some results of Chapter VIII are accepted on faith. The entire content of the book can be covered in a two-semester course.

The specialist will find here some novel approaches to familiar subjects and a substantial number of new proofs. In every field of mathematics a period of rapid growth leaves behind much disorder: various "folk theorems," as well as theorems with insufficient, sometimes even incorrect, proofs. I hope I have filled some of those gaps. As the line between pedantry and precision is thin, I might have crossed it in the wrong direction. However, in a field where an invocation of "it is easy to see" is sometimes considered a method of proof, there might be merit in actually writing the details out, messy as they might be.

This book originated in the course I gave for the first time at Berkeley in 1963 and a number of times, with a constantly changing content, at Rutgers and Bonn. Parts of it were written while I was a guest of the University of Bonn (Sonderforschungsbereich) and a member of the Institute for Advanced Study in Princeton. I am grateful to these institutions for providing me with excellent working conditions.

I am grateful to my colleagues, G. Bredon and P. Landweber, who read parts of the manuscript and contributed many useful suggestions. My wife never failed to emerge from her habitual location in the Middle Ages to provide much needed moral support.

Mrs. Louise Morse graciously provided the photograph serving as frontispiece. Taken in April of 1963, it shows all principal *dramatis personae* of this book, with the exception of R. Thom.

Notational Conventions

A cross-reference III,3.5 is to Theorem 5 of Section 3 of Chapter III; if the chapter number is omitted, it is to the chapter at hand. A reference A,3.5 refers to the Appendix following the last chapter.

The ring of integers is denoted by $\mathbf{Z}$, $n\mathbf{Z}$ stands for the subring of multiples of n, $\mathbf{Z}_n = \mathbf{Z}/n\mathbf{Z}$.

The Euclidean n-dimensional space is denoted by $\mathbf{R}^n$; when $\mathbf{R}^k$ is viewed as a subspace of $\mathbf{R}^n$, it is as the subspace of first k coordinates. The space of last k coordinates is denoted by $\bar{\mathbf{R}}^k$, $\mathbf{R}^n_+ = \{x = (x_1, \ldots, x_n) \in \mathbf{R}^n \,|\, x_n \geq 0\}$, $0 = (0, \ldots, 0) \in \mathbf{R}^n$. We write $D^n(r) = \{x \in \mathbf{R}^n \,|\, x^2 \leq r^2\}$, $\mathring{D}^n(r)$ for the interior of $D^n(r)$, and $S^n(r)$ for its boundary; if $r = 1$, then it is omitted from the notation.

The inverse of a matrix M is denoted M^{-1}, the transpose ${}^t M$. $\mathrm{Gl}(n)$ stands for the group of $n \times n$ nonsingular matrices; $\mathrm{Gl}(n)$ acts on $\mathbf{R}^n$ via the multiplication $v \mapsto M \cdot v$, $M \in \mathrm{Gl}(n)$, where we view the vector $v \in \mathbf{R}^n$ as an $n \times 1$ matrix, and the multiplication is, exceptionally, denoted by a dot. $V_{n,k}$ stands for the Stiefel manifold of k-frames in $\mathbf{R}^n$; this is the set of all $n \times k$ matrices of rank k. Sometimes we employ the same symbol for $n \times k$ matrices with orthonormal columns; the meaning is always clear from the context. $\mathbf{O}(n)$ stands for the group of orthogonal matrices; the inclusion $\mathbf{O}(n-1) \hookrightarrow \mathbf{O}(n)$ is given by

$$A \rightarrow \begin{pmatrix} 1 & 0 \\ 0 & A \end{pmatrix},$$

and the projection $\mathbf{O}(n) \rightarrow V_{n,k}$ by taking the first k columns.

We list now various groups defined in the text:

Γ^n, the group of diffeomorphisms of S^n modulo those which extend over D^n, III,6.

A^n, the group of invertible smooth structures on S^n, VI,2.

θ^n, the group of homotopy spheres, VIII,5.

P^n, the group of framed manifolds bounded by a homotopy sphere, X,6.

θ^n, the group of homotopy spheres, VIII,5.

P_0^n, the subgroup of P^n of manifolds with the boundary diffeomorphic to S^{n-1}, X,6.

$\Omega^k(M^m)$, the set of framed equivalence classes of closed framed k-dimensional submanifolds of M^m, IX,3.

$\Omega_{\Sigma}^k(M^m)$ (resp. $\Omega_S^k(M^m)$) the subset of $\Omega^k(M^m)$ consisting of these classes which can be represented by a framed homotopy sphere (resp. a framed sphere), IX,3.

Ω_f^k, the stable group $\Omega^k(S^m)$, m large, IX,6.2.

Σ_f^k, the subgroup of Ω_f^k consisting of framed homotopy spheres, X,6.3.

S_f^k, the subgroup of Ω_f^k consisting of framed spheres, IX,6.2.

$\pi_k(S)$, the stable group $\pi_{n+k}(S^n)$, n large, IX,6.2.

In IX,6.2, Ω_f^k is identified with $\pi_k(S)$; in IX,6.3 S_f^k is identified with the image of the stable Hopf–Whitehead homomorphism $J_k : \pi_k(\mathbf{SO}) \rightarrow \pi_k(S)$.

Differential Manifolds

I

Differentiable Structures

Les êtres de l'hyperspace sont susceptibles de définitions précises comme ceux
de l'espace ordinaire, et si nous ne pouvons nous les représenter, nous pouvons
les concevoir et les étudier.

H. Poincaré

1 Smooth Manifolds and Maps

A topological space is a manifold if it admits an open covering $\{U_\alpha\}$ where
each set U_α is homeomorphic, via some homeomorphism h_α, to an open
subset of the Euclidean space $\mathbf{R}^n$. A near-sighted topologist transferred from
$\mathbf{R}^n$ to a location in such a manifold would not notice a difference, at least
not until he or she decided to do Calculus: A function which is differentiable
when expressed in local coordinates relative to one U_α (i.e., its composition
with h_α) need not be differentiable relative to another set of local coordinates. For that to be true the covering must satisfy an additional condition.

(1.1) Definition Let M be a topological space. A *chart* in M consists of
an open subset $U \subset M$ and a homeomorphism h of U onto an open subset
of $\mathbf{R}^m$. A C^r *atlas* on M is a collection $\{U_\alpha, h_\alpha\}$ of charts such that the U_α
cover M and $h_\beta h_\alpha^{-1}$, the *transition maps*, are C^r maps on $h_\alpha(U_\alpha \cap U_\beta)$.

1

For our purpose, which is to define differentiable functions, two different atlases may yield the same result. They certainly will if they are compatible, in the sense that their union is an atlas. This relation of compatibility is an equivalence relation; hence every atlas is contained in a maximal one: the union of all atlases compatible with it. We now continue the Definition 1.1:

A maximal C^r atlas on M is called a C^r *structure*. A *differential manifold* M of class C^r consists of a second countable Hausdorff space M and a C^r structure on it.

A C^∞ atlas (chart, structure, . . .) will be referred to as smooth. Throughout this book we will consider smooth structures exclusively. This will not restrict the generality for it has been proved by H. Whitney [Wi2] that every C^r structure, r $\geq$ 1, contains a smooth structure.

A favorite method of studying smooth manifolds consists in observing how they are put together from smaller pieces. The pieces are not, however, smooth manifolds, they are *manifolds with boundary*. The definition of manifolds with boundary parallels 1.1; the only change is that we allow the h_α to be homeomorphisms onto open subsets of either $\mathbf{R}^m$ or $\mathbf{R}^m_+$, where $\mathbf{R}^m_+ = \{(x_1, \ldots, x_m) \in \mathbf{R}^m \mid x_m \geq 0\}$. The transition maps now become maps of open subsets of $\mathbf{R}^m_+$; such a map is, by definition, smooth if it is locally a restriction of a smooth map defined on an open subset of $\mathbf{R}^m$. The definitions of maximal atlas, smooth structure, etc., remain unchanged.

The set of those points of a smooth manifold which in a chart modeled on $\mathbf{R}^m_+$ correspond to points of $\mathbf{R}^{m-1}$ is topologically distinguished (think of local homology properties); it is denoted ∂M and called the *boundary* of M. Clearly, it is a closed subset of M. Its complement is called the *interior* of M and denoted Int M.

Now, let $\{U_\alpha, h_\alpha\}$ be an atlas on M. The topological invariance of ∂M implies that h_α maps $U_\alpha \cap \partial M$ to an open subset of $\mathbf{R}^{m-1}$. Thus $(U_\alpha \cap \partial M, h_\alpha \mid U_\alpha \cap \partial M)$ is a chart in ∂M. Moreover, the collection of all such charts is an atlas on ∂M. For if $h_\beta h_\alpha^{-1}$ is a transition map on $h_\alpha(U_\alpha \cap U_\beta)$ then $h_\beta h_\alpha^{-1}$ restricted to $h_\alpha(U_\alpha \cap U_\beta) \cap \mathbf{R}^{m-1}$ maps this set again to $\mathbf{R}^{m-1}$ and is smooth. Thus the boundary of an m-dimensional manifold M inherits from M a structure of a differential manifold without boundary and of dimension $m - 1$.

Throughout this book, and unless explicitly stated to the contrary, the word manifold will mean a smooth manifold with or without boundary. If $\partial M = \emptyset$, then we shall say that M is *closed*.

We will consider a few examples. Note that in order to specify a differential structure it is enough to describe one atlas. For instance, the *standard differential structure on* $\mathbf{R}^n$ refers to the atlas $\{U, h\}$ where $U = \mathbf{R}^n$, $h = \mathrm{id}$. Similarly, the complex n-dimensional space $\mathbf{C}^n$ becomes a smooth manifold via its identification with $\mathbf{R}^{2n}$.

(1.2) Let $M = S^n$, considered as the unit sphere in $\mathbf{R}^{n+1}$. Set $U_i^+ = \{(x_1, \ldots, x_{n+1}) \in S^n \,|\, x_i > 0\}$, $\quad U_i^- = \{(x_1, \ldots, x_{n+1}) \in S^n \,|\, x_i < 0\}$, $\quad i = 1, 2, \ldots, n + 1$, and let $h_i^\pm : U_i^\pm \to \mathbf{R}^n$ be given by leaving out the ith coordinate. This atlas defines the standard differential structure on S^n.

There is another way to define a differential structure on S^n: Let $a_\pm = (0, \ldots, 0, \pm 1) \in \mathbf{R}^{n+1}$, $U^\pm = S^n - \{a_\pm\}$ and let $h_\pm : U^\pm \to \mathbf{R}^n$ be the projection from $a_\pm$, i.e.,

$$h_\pm(x_1, \ldots, x_{n+1}) = \frac{1}{1 \mp x_{n+1}}(x_1, \ldots x_n, 0) \qquad \text{(see Fig. I.1)}.$$

Exercise Show that $\{U^\pm, h_\pm\}$ is an atlas compatible with the standard differential structure on S^n.

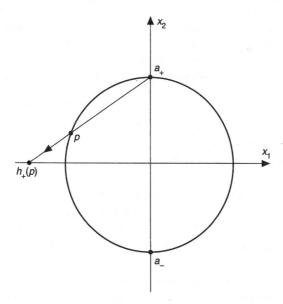

Figure I.1.

(1.3) The real projective n-space P^n is obtained by identifying antipodal points in S^n. Let $\pi: S^n \to P^n$ be the identification map. Note that π is a homeomorphism on every set U_i^+, $i = 1, \ldots, n+1$, and that their images cover P^n. Thus $\{\pi(U_i^+), h_i\pi^{-1}\}$ is an atlas on P^n.

To introduce a smooth structure on the complex projective space $\mathbf{C}P^n$ we use homogeneous coordinates. Namely, let V_i be the set of points in $\mathbf{C}P^n$ with the ith homogeneous coordinate $\neq 0$. Assigning to every point $(z_1, \ldots, z_{n+1})$ in V_i the point $(z_1/z_i, \ldots, z_{n+1}/z_i)$ with the ith coordinate omitted yields a homeomorphism h_i of V_i with $\mathbf{C}^n$. If $\mathbf{C}^n$ is identified with $\mathbf{R}^{2n}$, as before, then $\{V_i, h_i\}$ becomes an atlas on $\mathbf{C}P^n$.

(1.4) If $\{U_\alpha, h_\alpha\}$ is an atlas on M and $\{V_\beta, g_\beta\}$ is an atlas on N, and at least one of M, N is a closed manifold, then $\{U_\alpha \times V_\beta, h_\alpha \times g_\beta\}$ is an atlas on $M \times N$ and defines the product structure. But observe that if both M and N have non-empty boundaries, then this is not an atlas.

Exercise   Compute the transition maps in examples 1.2–1.4.

(1.5) A differential structure on a manifold M induces a differential structure on every open subset of M. In particular, writing the entries of an $n \times k$ matrix in succession identifies the set of all such matrices with $\mathbf{R}^{nk}$ and the subset $V_{n,k}$ of $n \times k$ matrices of maximal rank with an open subset of $\mathbf{R}^{nk}$. An $n \times k$ matrix of rank k can be viewed as a k-frame, that is, a set of k linearly independent vectors in $\mathbf{R}^n$; accordingly $V_{n,k}$, $k \leq n$, is called the Stiefel manifold of k-frames in $\mathbf{R}^n$. $V_{n,n}$ is, of course, the general linear group $\mathbf{Gl}(n)$. By the foregoing, $V_{n,k}$ is a differential manifold of dimension nk. (In II,2 we will define a differential structure on the group $\mathbf{O}(n)$ of orthogonal matrices.)

Now, we define the smooth maps.

(1.6) *Definition*   Let $f: M \to N$ where M, N are differential manifolds. We will say that f is *smooth* if there are atlases $\{U_\alpha, h_\alpha\}$ on M, $\{V_\beta, g_\beta\}$ on N, such that the maps $g_\beta f h_\alpha^{-1}$ are smooth wherever they are defined. We say that f is a *diffeomorphism* if it is smooth and has a smooth inverse.

The relation of diffeomorphism is an equivalence relation between smooth structures. That one is needed is illustrated by the following example. Let $h: \mathbf{R}^1 \to \mathbf{R}^1$ be a homeomorphism. Then $\{\mathbf{R}^1, h\}$ is an atlas on $\mathbf{R}^1$ and the smooth structures defined by two such atlases with homeomorphisms h_1, h_2 are distinct unless $h_1 h_2^{-1}$ and $h_2 h_1^{-1}$ are both differentiable.

Exercise   Show that all these structures are diffeomorphic.

Exercise   Let $T(\lambda, \mu) = \{x \in S^{\lambda+\mu-1} | \sum_{i\le\lambda} x_i^2 = \sum_{i>\lambda} x_i^2\}$, $D(\lambda, \mu) = \{x \in S^{\lambda+\mu-1} | \sum_{i\le\lambda} x_i^2 \ge \sum_{i>\lambda} x_i^2\}$, where $S^{\lambda+\mu-1}$ is the unit sphere in $\mathbf{R}^{\lambda+\mu}$.
   Show that $T(\lambda, \mu)$ is diffeomorphic to $S^{\lambda-1} \times S^{\mu-1}$ and $D(\lambda, \mu)$ is diffeomorphic to $D^\lambda \times S^{\mu-1}$.

It would have been possible to define, in an obvious way, C^r diffeomorphisms, $r > 0$. But, again, there is no restriction in generality in considering only smooth, i.e., C^∞, diffeomorphisms: H. Whitney has shown that if two smooth manifolds are C^r diffeomorphic, $r > 0$, then they are C^∞ diffeomorphic.

(1.7)   Consider the manifold $\mathbf{Gl}(n)$. If $A, B \in \mathbf{Gl}(n)$, then the entries of the matrix AB are polynomials in terms of entries of A and B. Thus the group operation in $\mathbf{Gl}(n)$ defines a smooth map $\mathbf{Gl}(n) \times \mathbf{Gl}(n) \to \mathbf{Gl}(n)$. A group with these properties, i.e., such that the underlying space is a smooth manifold and the group operation a smooth map, is called a Lie group. We will show in II,2 that $\mathbf{O}(n)$ is also a Lie group.
   Note that if G is a Lie group and $g_0 \in G$, then the map of G into itself given by $g \mapsto g_0 g$ is a diffeomorphism.

(1.8)   Let $\mathscr{U} = \{U_\alpha, h_\alpha\}$ be an atlas in M. We assume that an orientation of $\mathbf{R}^n$ has been chosen once and for all. Then, every chart (U_α, h_α) determines a local orientation in U_α, i.e., a preferred generator of $H^n(M, M - p)$ at every point $p \in U_\alpha \cap \text{Int } M$ (cf. [D,VIII,2.1]). Two such orientations in U_α and U_β are compatible if and only if the determinant of the Jacobian of the corresponding transition map is everywhere positive. An atlas on M with this property is called *oriented* and a maximal oriented atlas is called an *oriented smooth structure* on M.
   If $\{U_\alpha, h_\alpha\}$ and $\{V_\beta, g_\beta\}$ are oriented atlases on M and N respectively, then a diffeomorphism $f : M \to N$ is said to be *orientation preserving* if the Jacobians of all maps $g_\beta f h_\alpha^{-1}$ have positive determinant.
   If M has a non-empty boundary and $\mathscr{U} = \{U_\alpha, h_\alpha\}$ is an oriented smooth structure on M, then the smooth structure on ∂M induced by $\mathscr{U}$ is also oriented. For if $p \in \partial M \cap U_\alpha \cap U_\beta$ and $h_\beta h_\alpha^{-1} = (f_1, \ldots, f_n)$, then the

Jacobian J of $h_\beta h_\alpha^{-1}$ at $h_\alpha^{-1}(p)$ has the form

$$\begin{pmatrix} J' & * \\ 0 & \partial f_n/\partial x_n \end{pmatrix},$$

where J' is the Jacobian of $h_\beta h_\alpha^{-1}$ restricted to ∂M and $\partial f_n/\partial x_n > 0$. Since $\det J > 0$, $\det J' > 0$.

Exercise Show that if S^n is oriented as the boundary of D^{n+1}, then the map h_+ in 1.2 is orientation preserving.

2 Partitions of Unity

Many constructions in differential topology are performed with the help of partitions of unity. For this purpose it will be useful to have a special kind of atlas available:

(2.1) *Definition* An atlas $\{U_\alpha, h_\alpha\}$ on M is said to be *adequate* if it is locally finite, $h_\alpha(U_\alpha) = \mathbf{R}^m$ or $\mathbf{R}_+^m$ and $\bigcup_\alpha h_\alpha^{-1}(\mathring{D}^m) = M$.

(A family of subsets of M is locally finite if every point of M is contained in an open neighborhood intersecting at most a finite number of them. Recall that a space M is said to be paracompact if every open covering admits a locally finite refinement.)

(2.2) **Theorem** *Let $\mathcal{V} = \{V_\beta\}$ be a covering of M. Then there is an adequate atlas $\{U_\alpha, h_\alpha\}$ such that $\{U_\alpha\}$ is a refinement of $\mathcal{V}$.*

In particular, it follows that a smooth manifold is paracompact.

Proof Since M is locally compact, Hausdorff, and second countable, we obtain easily (*cf.* [Du,XI,7.2]) that there is a sequence $\{K_i\}$, $i = 1, 2, \ldots$, of open subspaces of M, with compact closures, and such that $\bar{K}_i \subset K_{i+1}$ and $\bigcup_i K_i = M$. We also set $K_0 = K_{-1} = \emptyset$.

We construct now the desired refinement in stages; the ith stage is as follows: Let $p \in \bar{K}_i - K_{i-1}$ and suppose that $p \in V_\beta$. Let (U_p, h_p) be a chart such that $h_p(U_p) = \mathbf{R}^n$, $h_p(p) = \mathbf{0}$, and $U_p \subset (K_{i+1} - \bar{K}_{i-2}) \cap V_\beta$.

Now, $\bar{K}_i - K_{i-1}$ is compact, and the sets $h_p^{-1}(\mathring{D}^m)$ cover it. Hence there is a finite family that does the same; let $U_1^i, \ldots, U_{k_i}^i$ be the corresponding charts. Then the family $\mathcal{U} = \{U_j^i\}$, $i = 1, 2 \ldots, j = 1, 2, \ldots, k_i$ is locally finite: Every point of M is contained in one of the open sets K_i and each such set intersects—at most—only the U_j^m with $m \leq i + 2$. It is clear that $\mathcal{U}$ is a refinement of $\mathcal{V}$ and an adequate atlas.   □

Now, let $\{U_\alpha, h_\alpha\}$ be an adequate atlas on M. Let λ be a smooth nonnegative function on $\mathbf{R}^m$ which equals 1 on D^m and zero outside of $D^m(2)$. Let $\lambda_\alpha = \lambda h_\alpha$ in U_α and 0 outside U_α; we will call the family $\{\lambda_\alpha\}$ the family of bump functions associated to $\{U_\alpha\}$. Let $\mu_\alpha(p) = \lambda_\alpha(p)/\sum_\alpha \lambda_\alpha(p)$. The family $\{\mu_\alpha\}$ is the *partition of unity* associated to $\{U_\alpha, h_\alpha\}$.

We will apply partitions of unity to some extension and approximation problems on smooth manifolds.

(2.3) *Definition* Let K be a subset of a smooth manifold M and f a map of some subset of M containing K to a smooth manifold N. We say that f is *smooth on K* if its restriction to K is locally a restriction of a smooth map, i.e., if for every point $p \in K$ there is an open neighborhood U of p in M and a smooth map $F: U \to N$ that agrees with f on $U \cap K$.

In general, a smooth function on K need not be a restriction of a smooth function on M. However, this is the case if K is a closed subset:

(2.4) *Theorem* Let K be a closed subset of M and $f: K \to \mathbf{R}$ a smooth function. Then f is a restriction of a smooth function on M.

Proof For every point $p \in K$ let U_p, $F_p: U_p \to \mathbf{R}$ be as in 2.3. Consider the covering of M consisting of sets U_p and $M - K$ and let $\{V_\alpha\}$ be an adequate atlas refining it, and $\{\mu_\alpha\}$ an associated partition of unity. Define functions g_α on M as follows: If V_α is contained in one of the sets U_p then set $g_\alpha = \mu_\alpha F_p$; otherwise $g_\alpha = 0$. Now let $g(p) = \sum_\alpha g_\alpha(p)$. For every point p this sum is actually finite on a neighborhood of p; hence g is a smooth function on M. If $q \in K$, then $g_\alpha(q)$ is either zero or $\mu_\alpha(q)F_p(q)$, that is, $\mu_\alpha(q)f(q)$. Hence

$$g(q) = \sum_\alpha \mu_\alpha(q)f(q) = f(q) \sum_\alpha \mu_\alpha(q) = f(q).   \square$$

Exercise Show that a smooth manifold is a normal topological space. (It is known that every paracompact space is normal [Du,VIII,2.2].)

Now, we have an approximation theorem.

(2.5)   Theorem   *Let $f: M \to \mathbf{R}^n$ be a continuous map, smooth on a closed subset K of M, and let $\varepsilon > 0$. Then there is a smooth map $g: M \to \mathbf{R}^n$ that agrees with f on K and such that $|f(p) - g(p)| < \varepsilon$.*

Proof   For every $p \in K$ there is a neighborhood U_p and a smooth map $f_p: U_p \to \mathbf{R}^n$ that agrees with f on K. Choosing U_p small enough we can also achieve that $|f_p(x) - f(x)| < \varepsilon$ for all x in U_p. If p is not in K, then there is a neighborhood U_p of p disjoint from K and such that $|f(p) - f(x)| < \varepsilon$ for all x in U_p. Let $\{V_\alpha\}$ be an adequate atlas refining $\{U_p\}$ and $\{\mu_\alpha\}$ an associated partition of unity. For each V_α choose a U_p containing it and define $g_\alpha: M \to \mathbf{R}^n$ by setting $g_\alpha = f_p$ if the chosen U_p intersects K; $g_\alpha = f(p)$ otherwise. In both cases we have $|g_\alpha(x) - f(x)| < \varepsilon$ for all x in V_α.

Now, let $g = \sum_\alpha \mu_\alpha g_\alpha$. As in the proof of 2.4 we see that g is a smooth map agreeing with f on K. We have

$$|f(x) - g(x)| = \left| \sum_\alpha \mu_\alpha(x) f(x) - \sum_\alpha \mu_\alpha(x) g_\alpha(x) \right|$$

$$\leq \sum_\alpha \mu_\alpha(x) |f(x) - g_\alpha(x)| < \sum_\alpha \mu_\alpha(x) \varepsilon = \varepsilon. \qquad \square$$

Note that with a minimal change in the proof the constant ε can be replaced by a non-negative continuous function on M. It is less trivial, but true, that 2.5 is valid with $\mathbf{R}^n$ replaced by an arbitrary manifold N with some metric on it. This will be shown in III,2.5.

(2.6)   Corollary   *If M is connected, then every two points of M can be joined by a smooth curve.*

Proof   There certainly is a piecewise smooth curve joining two given points. Now, every corner of such a curve lies in one chart and 2.5 is used to smooth it. $\square$

3   Smooth Vector Bundles

There are two elements in the idea of a vector bundle: the local product structure, and the algebraic operation in the fibers. In the case of a vector bundle over a smooth manifold we have naturally the notion of a smooth vector bundle: one in which both of these are smooth.

To make this precise, we establish the notation first; we assume that the reader is familiar with the elements of the theory of vector bundles, e.g., [MS, § 2–3].

Let $\pi\colon E \to M$ be an n-dimensional vector bundle over a smooth manifold M and let $\{U_\alpha, h_\alpha\}$ be an atlas on M such that the bundle is trivial over each of the sets U_α. (We will sometimes say that such charts are trivializing for E.) Let ϕ_α be the composition of the canonical map $\pi^{-1}(U_\alpha) \to U_\alpha \times \mathbf{R}^n$ with the projection on $\mathbf{R}^n$. Then (ϕ_α, h_α) sends $\pi^{-1}(U_\alpha)$ homeomorphically onto an open subset of $\mathbf{R}^n \times \mathbf{R}^m$ (or $\mathbf{R}^n \times \mathbf{R}^m_+$); if these maps form a smooth atlas on E, then we say that E is a smooth vector bundle.

There is an equivalent definition, which is more convenient to use. Let $\phi_{\alpha, p}\colon \mathbf{R}^n \to \pi^{-1}(p)$ be the right inverse of ϕ_α. If $p \in U_\alpha \cap U_\beta$, then $\phi_\beta \phi_{\alpha, p}$ is an isomorphism $\mathbf{R}^n \to \mathbf{R}^n$. Viewing it as an element of $\mathrm{Gl}(n)$, we obtain a map $\Phi_{\alpha,\beta}\colon U_\alpha \cap U_\beta \to \mathrm{Gl}(n)$, $\Phi_{\alpha,\beta}(p) = \phi_\beta \phi_{\alpha, p}$. Clearly:

(3.1) Proposition *E is a smooth vector bundle if and only if all maps $\Phi_{\alpha,\beta}$ are smooth.* $\square$

Exercise Show that the dual of a smooth vector bundle is a smooth bundle, and that the Whitney sum of two smooth vector bundles is a smooth vector bundle.

$\mathbf{R}^n$ is endowed with an inner product, but there is no *a priori* reason for the maps $\phi_\beta \phi_{\alpha, p}$ to preserve it. This means that we cannot use the maps $\phi_{\alpha, p}$ to define a smooth inner product in every fiber of E. The next two theorems will show that this is nevertheless possible, and in essentially one way only.

The structure we are going to discuss is called a *Riemannian metric,* r-metric for short. To define a smooth r-metric, take the Whitney sum F of the bundle E with itself, and let $\sigma\colon F \to M$ be the projection of its total space onto M. F is again a smooth bundle; we will consider the vectors in F to be pairs of vectors from the same fiber of E.

Now, a smooth r-metric on E is a smooth map $F \to \mathbf{R}$ which on every fiber is a symmetric, positive definite bilinear form.

(3.2) Theorem *A smooth vector bundle $\pi\colon E \to M$ admits a smooth Riemannian metric.*

Proof Let $\{U_\alpha\}, \{\phi_\alpha\}$ be as before; by 2.2 we can assume that $\{U_\alpha\}$ is an adequate atlas; let $\{\mu_\alpha\}$ be an associated partition of unity. Now, choose a symmetric, positive definite bilinear form $\Omega(\ ,\ )$ on $\mathbf{R}^n$ and define $\langle\ ,\ \rangle_\alpha\colon \sigma^{-1}(U_\alpha) \to \mathbf{R}$ by $\langle v, w\rangle_\alpha = \Omega(\phi_\alpha(v), \phi_\alpha(w))$; this is a smooth map. Let $\langle\ ,\ \rangle\colon F \to \mathbf{R}$ be given by

$$\langle v, w\rangle = \sum_\alpha \mu_\alpha(\pi(v))\langle v, w\rangle_\alpha.$$

(If $\pi(v)$ is not in U_α then $\langle v, w\rangle_\alpha$ is not defined, but then $\mu_\alpha(\pi(v)) = 0$ as well, so this does not matter.)

Now, $\langle\ ,\ \rangle$ is certainly a smooth map; to show that it is an r-metric, notice that on each fiber $\langle\ ,\ \rangle$ is a linear combination of symmetric, positive definite bilinear forms. Thus it is certainly symmetric and bilinear itself. Since the coefficients of this linear combination are non-negative and not all zero (their sum equals 1), it is also positive definite.   □

Let E be a vector bundle with a smooth r-metric. Fix an orthonormal basis $\mathbf{b}$ in $\mathbf{R}^n$ and let U_α be a trivializing chart. Then $\mathbf{b}_p = \phi_{\alpha, p}(\mathbf{b})$ is a basis in the fiber over p and all of these form a smooth family of bases over $\pi^{-1}(U_\alpha)$. Using the Gram–Schmidt orthogonalization procedure (*cf.* A,4.1) with respect to the inner product in each fiber we can obtain a family $\mathbf{b}'_p$ of orthonormal bases in every fiber over U_α. This in turn can be used to define new canonical maps ϕ_α by the requirement that $\phi_\alpha(\mathbf{b}'_p) = \mathbf{b}$. The maps $\Phi_{\alpha,\beta}$ that we get from these are smooth and map $U_\alpha \cap U_\beta$ to the subgroup $\mathbf{O}(n)$ of $\mathbf{Gl}(n)$. This means that the introduction of an r-metric allows the reduction of the group of the bundle to $\mathbf{O}(n)$.

We can now state the uniqueness theorem for r-metrics.

(3.3) Theorem *Let $f\colon E \to F$ be an isomorphism between two smooth r-bundles. Then there is an isometry $g\colon E \to F$.*

Actually, we will show that there is a smooth family F_t, $t \in I$, of isomorphisms such that $F_0 = f$, $F_1 = g$. We will return to this point in II,4.6.

Proof We suppose we are given two r-metrics on the same bundle E; the general case clearly reduces to this. As we have just seen, given a trivializing neighborhood U, these metrics can be represented by two families $\mathbf{b}, \mathbf{b}'$ of bases, each orthonormal relative to the corresponding metric. An automorphism of E is represented in U by a family of nonsingular matrices

$M_p, p \in U$, each M_p giving the coordinates in terms of $\mathbf{b}'_p$ of the transformed vectors of the base $\mathbf{b}_p$. Such an automorphism is an isomorphism if and only if all of the M_p are orthogonal.

To prove the theorem, we start with the identity automorphism of E and suppose that in a trivializing neighborhood U it is given by matrices $M_p, p \in U$. By A,4.3 we can write $M_p = O_p S_p$, where O_p is orthogonal, S_p is symmetric positive definite, and both depend smoothly on p. Then the matrices O_p define an isometry $\theta_U \colon E \,|\, U \to E \,|\, U$.

We claim that $\theta_U = \theta_V$ in $U \cap V$, that is, the isomorphism represented by O_p does not depend on the choice of orthonormal bases $\mathbf{b}$, $\mathbf{b}'$. This is so, for changing the bases amounts to replacing M_p by $N M_p K$, with $N, K \in \mathbf{O}(n)$. Since $N M_p K = (N O_p K)(K^{-1} S_p K)$ with $N O_p K \in \mathbf{O}(n)$ and $K^{-1} S_p K$ symmetric positive definite, the uniqueness part of A,4.3 implies that $N O_p K$ is the orthogonal matrix in the representation of $N M_p K$ as a product. But $N O_p K$ is the representation of the same isometry with respect to the new bases.

We have shown that the family of isometries θ_U represents a well-defined isometry of the bundle E. The same argument applied to matrices S_p shows that they define an automorphism of E.

To obtain the family F_t of automorphisms consider $M_p(t) = O_p(t I_n + (1 - t) S_p)$. Since the set of positive definite symmetric matrices is convex, the same argument as before shows that this does define the desired family of automorphisms. $\quad\square$

Given a vector bundle with an r-metric $\langle \ , \ \rangle$, we define the length $|v|$ of a vector v to be $\langle v, v \rangle^{1/2}$ and the distance between two vectors in the same fiber as the length of their difference. Theorem 2.5 generalizes easily to cross sections of a vector bundle:

(3.4) Theorem *Let $s \colon M \to E$ be a continuous section of a smooth vector bundle with an r-metric. Suppose that s is smooth on a closed subset K of M and let $\varepsilon > 0$. Then there is a smooth section $t \colon M \to E$ that agrees with s on K and such that $|s(p) - t(p)| < \varepsilon$ for all p in M.*

Proof By 2.5 the theorem is true if E is a trivial bundle. So choose an adequate atlas $\{U_\alpha\}$ on M such that the bundle is trivial over every U_α, for each α find a section t_α over U_α ε-approximating s, and glue all those sections together using the associated partition of unity: $t = \sum_\alpha \mu_\alpha t_\alpha$. $\quad\square$

(3.5) A Riemannian metric provides a convenient way to describe an operation of "shrinking" a vector bundle. Let ε be a smooth positive function on M and E the total space of a smooth bundle over M with projection π. Consider the map $F\colon E \to E$ given by

$$F(v) = \varepsilon(\pi(v)) \frac{v}{(1 + v^2)^{1/2}}.$$

F maps smoothly the fiber over p onto an open disc in it of diameter $\varepsilon(p)$; thus $F(E)$ is an open disc bundle. Since F is a diffeomorphism, the bundle structure on E induces a smooth vector bundle structure on $F(E)$. We will call this operation ε-*shrinking* of E.

(3.6) Recall that a bundle E is oriented if all transition maps $\Phi_{\alpha,\beta}$ are maps to the subgroup $\mathrm{Gl}_+(n)$ of matrices with positive determinant. The argument given in 3.2 shows at the same time that the group of an oriented bundle can be reduced to $\mathrm{SO}(n)$.

If E is oriented and $g_n \in H_n(\mathbf{R}^n, \mathbf{R}^n - \mathbf{0})$ is the canonical orientation of $\mathbf{R}^n$, then $(\phi_{\alpha,p})_* g_n \in H_n(E_p, E_p - \mathbf{0})$ is an orientation of E_p that does not depend on α. Thus, all fibers of an oriented bundle are canonically oriented.

4 Tangent Space

The notion of a tangent space to a surface is quite intuitive, but the intuition depends strongly on the fact that a surface is a submanifold of $\mathbf{R}^3$. Nevertheless, it is possible to define the tangent space using only the smooth structure. We will do this now and show that the union of all tangent spaces forms a smooth vector bundle, the tangent bundle. The merit of proceeding this way is that the tangent bundle emerges as an invariant of the smooth structure.

A vector at a point is "a direction and a magnitude." It is possible to translate this idea into the context of charts: the direction at a point would be a suitably defined equivalence class of smooth curves. But it is easier to adopt an "operational" point of view: A vector at a point associates to every function defined in the neighborhood a number: its derivative in the direction of the vector. This operation has certain formal properties and our point of view will be to identify vectors at a point with an operation having these properties.

Let p be a point of a smooth m-dimensional manifold M.

(4.1) *Definition* A *tangent vector* X at p is an operation which associates a number Xf to every smooth function f defined in a neighborhood of p, and satisfies the following conditions:

(a) If f and g agree in a neighborhood of p, then $Xf = Xg$;
(b) $X(\lambda f + \mu g) = \lambda Xf + \mu Xg$ for every two numbers λ, μ;
(c) $X(fg) = (Xf)g(p) + f(p)(Xg)$.

The set of all tangent vectors at p will be denoted T_pM, or simply T_p if appropriate.

As an example, let (U, h) be a chart at $p \in \text{Int } M$, where $h(q) = (x_1(q), \ldots, x_m(q)) \in \mathbf{R}^m$, $q \in U$. Then associating to every function f the partial derivative $\partial fh^{-1}/\partial x_i$ at $h(p)$ we obtain a vector in T_p. It will be denoted $\partial/\partial x_i$ or ∂_i if there is no danger of confusion.

For instance, if $M = \mathbf{R}, (U, h) = (\mathbf{R}, \text{id})$, we get for every $t_0 \in \mathbf{R}$ a vector in T_{t_0}, which will be denoted ∂t.

Observe that the same construction applies if $p \in \partial M$. For although fh^{-1} is defined only in a neighborhood of $h(p)$ in $\mathbf{R}^m_+$, it extends over a neighborhood in $\mathbf{R}^m$ and the partial derivatives at $h(p)$ do not depend on the choice of the extension.

It follows immediately from (b) and (c) that if f is constant in a neighborhood of p then $Xf = 0$.

Our task now will be to show that each T_p is an m-dimensional vector space and their union a smooth vector bundle over M.

Clearly, with the obvious definition of multiplication and addition, T_p is a vector space. We will show that $\partial_1, \ldots, \partial_m$ form a basis.

Let then f be a function defined in a chart (U, h) and consider $g = fh^{-1}$. We will assume that $f(p) = 0$ and that $h(p) = \mathbf{0}$. By A,2.1, in some neighborhood of $\mathbf{0}$ we have $g = \sum_i x_i g_i$, where $g_i(\mathbf{0}) = (\partial g/\partial x_i)(\mathbf{0})$. (Again, this makes sense even when p is in ∂M.) Thus $f(q) = gh(q) = \sum_i x_i(q)g_i(h(q))$, and if x is a vector at p, then

(4.1.1) $\quad Xf = X\left(\sum_i x_i g_i(h)\right) = \sum_i (Xx_i)g_i(h(p)) = \sum_i (Xx_i)\,\partial_i f,$

since $x_i(p) = 0$ and $g_i(h(p)) = g_i(\mathbf{0}) = (\partial fh^{-1}/\partial x_i)(\mathbf{0})$. This shows that X is a linear combination of ∂_i, and since the ∂_i are linearly independent $(\partial_i(x_j) = \delta_{ij})$ they form a basis.

The assumption that $f(p) = 0$ is inessential: Setting $f' = f - f(p)$ we have $Xf = Xf'$ and $f'(p) = 0$. Finally, if $h(p) \neq \mathbf{0}$ then 4.1.1 is still valid with $\partial_i f$ standing for the derivatives of fh^{-1} at $h(p)$. Collecting all this, we have:

(4.2) **Proposition** *Let (U, h) be a chart in M such that $h(U)$ is convex. Let $p \in U$ and $X \in T_p$. Then $X = \sum_i \alpha_i \partial_i$, where $\alpha_i = Xx_i$ and $\partial_i f = \partial fh^{-1}/\partial x_i$ at $h(p)$.* $\square$

We will call $\alpha_1, \ldots, \alpha_m$ the coordinates of X with respect to the chart U.

Suppose now that M, N are two smooth manifolds, $f: M \to N$ a smooth map, $p \in M$ and $X \in T_p M$. We define a vector $Y \in T_{f(p)} N$ by requiring that $Yg = X(gf)$ for every smooth function g in a neighborhood of $f(p)$. It is a routine task to check that Y is, indeed, a vector at $f(p)$ and that setting $Y = Df_p X$ we obtain a homomorphism $Df_p: T_p M \to T_{f(p)} N$. We want to find the coordinates of Y. Let then (U_α, h_α) and (U_β, h_β) be two charts, about p and $f(p)$ respectively, let $X_\alpha = {}^t(\alpha_1, \ldots, \alpha_m)$ be an $m \times 1$ matrix whose coefficients are the coordinates of X rel. U_α, and similarly $Y_\beta = {}^t(\beta_1, \ldots, \beta_n)$. Setting $h_\alpha = (x_1, \ldots, x_m)$, $h_\beta = (y_1, \ldots, y_n)$ we have, by 4.1.1,

$$\beta_j = Y(y_j) = X(y_j(f)) = \sum_i \alpha_i \frac{\partial}{\partial x_i} y_j(fh_\alpha^{-1}),$$

i.e.,

(4.2.1) $Y_\beta = J_\beta^\alpha(f, p) \cdot X_\alpha$ (matrix multiplication),

where $J_\beta^\alpha(f, p)$ is the Jacobian matrix of $h_\beta fh_\alpha^{-1}$ at $h_\alpha(p)$.

Let $TM = \bigcup_{p \in M} T_p M$ and let $\pi: TM \to M$ be given by $\pi(T_p) = p$. Fix a chart U. By 4.2 the assignment

$$X \mapsto (p, \alpha_1, \ldots, \alpha_m),$$

where $p = \pi(X)$ and the α_i are the coordinates of X rel. U, defines a one-to-one map of $\pi^{-1} U$ onto $U \times \mathbf{R}^m$. We topologize TM by requiring these maps to be homeomorphisms.

If $f: M \to N$, then we define $Df: TM \to TN$, the *differential of f*, by $Df | T_p M = Df_p$. We have the following fundamental theorem:

(4.3) **Theorem** *The projection π gives TM the structure of a smooth vector bundle over M, the tangent bundle of M. Df is a bundle morphism and a smooth map of TM to TN.*

Proof To see that TM is a smooth vector bundle look at 4.2.1, in which we set $M = N, f = \text{id}$. Now, the map $p \mapsto J_\beta^\alpha(\text{id}, p)$ is a smooth map into $\mathbf{Gl}(m)$ and it is precisely the map $\Phi_{\alpha, \beta}$ from 3.1, whence the smoothness of TM.

We have already noted that Df is linear on each fiber of TM; that it is smooth follows from 4.2.1 and the fact that the map $p \mapsto J_\beta^\alpha(f, p)$ is a smooth map of $h_\alpha f^{-1}(U_\beta)$ to the space of $m \times n$ matrices. ☐

The dual bundle to the tangent bundle will be denoted T^*M and called the *cotangent bundle*, and its elements cotangent vectors. It is again a smooth bundle (*cf.* exercise in 3.1). We coordinatize it as follows.

If (U, h) is a chart, $h = (x_1, \ldots, x_m)$ and $q \in U$, then the covectors $dx_i \in T_q^*M$, $i = 1, \ldots, m$, are defined by

$$dx_i(X) = X(x_i), \qquad X \in T_qM.$$

Since $dx_i(\partial_j) = \delta_{ij}$, the dx_i form a basis.

The cotangent bundle is isomorphic to the tangent bundle, but non-canonically. A specific isomorphism is given if the tangent bundle is endowed with an r-metric. If this is the case, then one can associate to every vector $X \in T_pM$ the covector $L_X \in T_p^*M$ given by the rule

$$L_X(Y) = \langle X, Y \rangle.$$

This defines a map $L\colon TM \to T^*M$.

(4.4) Proposition *L is a smooth isomorphism.*

Proof That L is an isomorphism on every fiber is a standard proposition from linear algebra. That it is smooth follows from the fact that finding coordinates of L_X amounts to solving a system of linear equations with the matrix of coefficients of rank m and with smooth entries. ☐

(4.5) A simple case of 4.2.1 is when M or N is the line $\mathbf{R}$. $T\mathbf{R}$ is a 1-dimensional bundle and we can take as the base at every point the vector ∂t. Now, if $f\colon M \to \mathbf{R}$ and $X \in TM$, then $Df(X) = (Xf)\,\partial t$, or, simply

(4.5.1) $Df(X) = Xf.$

In particular, if $f\colon \mathbf{R} \to \mathbf{R}$ then $Df(\partial t) = df/dt$.

If $f\colon \mathbf{R} \to M$, then the vector $Df_{t_0}(\partial t)$ will be called the vector tangent to the curve f at $t = t_0$.

Let $v \in \mathbf{R}^m$ and $f_v\colon \mathbf{R} \to \mathbf{R}^m$ be given by $f_v(t) = tv$. It follows from 4.2.1 that by associating to v the vector tangent to f_v at $\mathbf{0}$ one obtains an isomorphism $\mathbf{R}^m \to T_0\mathbf{R}^m$. Its inverse will be called the *exponential map* and denoted exp. Clearly, exp associates to a vector in $T_0\mathbf{R}^m$ the vector in $\mathbf{R}^m$

with the same coordinates—assuming of course that we take as coordinates in $T_0 \mathbf{R}^m$ those induced by the chart $(\mathbf{R}^m, \text{id})$.

We can define this map at every $v \in \mathbf{R}^m$: first, use the differential of the translation $x \mapsto x - v$ to identify $T_v \mathbf{R}^m$ with $T_0 \mathbf{R}^m$, and then set

$$\exp_v(w) = \exp(w) + v.$$

This yields a smooth map of $T\mathbf{R}^m$ to $\mathbf{R}^m$, which is a diffeomorphism on every fiber and on the zero section. In Chapter III the exponential map will be defined for all manifolds.

Suppose that M has a non-empty boundary and consider the differential of the inclusion $\partial M \hookrightarrow M$. At every point $p \in \partial M$ it is a monomorphism; thus the image of $T_p(\partial M)$ is a well-defined $(m-1)$-dimensional subspace of $T_p M$. It follows from 4.2.1 that, independently of the chart chosen, the vectors in $T_p(\partial M)$ viewed as vectors in $T_p M$ have the last coordinate equal 0.

$T_p(\partial M)$ divides $T_p M$ into two half-spaces; it is possible to distinguish between them geometrically: Let $c(t)$ be a curve in M beginning at p, i.e., a map $c: \mathbf{R}_+ \to M$, $c(0) = p$. It is easy to see, again by 4.2.1, that the tangent vector to c at p has the last coordinate non-negative. This justifies saying that a vector in $T_p M$ points inside M if it has the last coordinate positive. Observe now that, for any $t \in \mathbf{R}_+$, if $c(t) \in \partial M$ then the tangent vector to c at $c(t)$ either points inside M or is in the tangent space to ∂M.

Exercise   Consider $S^n \subset \mathbf{R}^{n+1}$. If $v \in S^n$, then $T_v S^n \subset T_v \mathbf{R}^{n+1}$. Show that $\exp_v(T_v S^n)$ is the n-dimensional plane in $\mathbf{R}^{n+1}$ perpendicular to v.

Exercise   Show that the tangent bundle to an oriented manifold is oriented.

Exercise   Show that the tangent bundle to a Lie group is trivial.

5   Vector Fields

Since TM is a smooth manifold it makes sense to talk of smooth sections:

(5.1)   *Definition*   A *vector field* on a manifold M is a smooth map $X: M \to TM$ such that $\pi X = \text{id}$.

If X is a vector field we will write $X(p) = X_p$.

Let (U, h) be a chart in M and X a vector field. Since $X_p \in T_pM$, we have, by 4.2,

(5.1.1) $$X_p = \sum_i \alpha_i(p) \, \partial_i.$$

Since X is smooth the functions α_i must be smooth. Conversely, if $X: M \to TM$ is a section and 5.1.1 holds in every chart for some smooth functions α_i, then, clearly, X is a vector field. This implies that if X, Y are vector fields and λ, μ real numbers, then $\lambda X + \mu Y$ (defined by $(\lambda X + \mu Y)_p = \lambda X_p + \mu Y_p$) is a vector field. In the same way, one shows that if f is a smooth function on M, then fX, defined by $(fX)_p = f(p)X_p$, is a vector field.

Let $\mathscr{X}(M)$ denote the set of all vector fields on M, $C^\infty(M)$ the set of all smooth functions on M. We have introduced in $\mathscr{X}(M)$ the operations of addition and of multiplication by elements of $C^\infty(M)$. A routine verification yields the following:

(5.2) Proposition $\mathscr{X}(M)$ *is a vector space and a module over* $C^\infty(M)$. □

We can make $\mathscr{X}(M)$ operate on $C^\infty(M)$: If $X \in \mathscr{X}(M)$ and $f \in C^\infty(M)$, then we define the function Xf by $(Xf)_p = X_p f$. To show that Xf is smooth, note that, by 5.1.1,

$$Xf(p) = \sum_i \alpha_i(p) \frac{\partial f h^{-1}}{\partial x_i}$$

in some chart (U, h). The same formula shows that if λ, μ are numbers then $X(\lambda f + \mu g) = \lambda(Xf) + \mu(Xg)$, i.e., that X induces an automorphism of $C^\infty(M)$ considered as a vector space.

If we apply X to a product of functions, then by 4.1(c) we get

(5.2.1) $$X(fg) = (Xf)g + f(Xg).$$

It is interesting to note that 5.2.1 characterizes vector fields:

(5.3) Proposition *An endomorphism X of $C^\infty(M)$ satisfying* 5.2.1 *is a vector field.*

Proof Let X satisfy 5.2.1 and let f be a smooth function in a neighborhood of $p \in M$. Define X_p by $X_p(f) = (Xf)(p)$. Then X_p satisfies 4.1(b), (c), and to show that it satisfies (a) it is enough to show that if f vanishes in a neighborhood of p, then $X_p(f) = 0$. This follows, for there is a smooth function λ on M such that $\lambda f = f$ and $\lambda(p) = 0$; hence $X_p(f) = X_p(\lambda f) = X_p(\lambda)f(p) + \lambda(p)X_p(f) = 0$. $\square$

If X and Y are two vector fields, then XY denotes the endomorphism Y followed by X. We define the bracket $[X, Y]$ by the formula

$$[X, Y] = XY - YX.$$

A routine verification shows that $[X, Y]$ satisfies 5.2.1; hence,

(5.4)   Corollary $[X, Y]$ *is a vector field.* $\square$

We will say more about the bracket of vector fields in Chapter V.

A smooth section of the cotangent bundle is called a *covector* field. A smooth function $f: M \to \mathbf{R}$ gives rise to a covector field df defined by

$$df(X_p) = X_p f, \qquad X_p \in T_p M.$$

A calculation shows that, in terms of a chart (U, h),

$$df = \sum_i \frac{\partial f h^{-1}}{\partial x_i} \, dx_i;$$

hence df is indeed a covector field. It is called the differential of f. (This terminology confuses df and Df. This is traditional and not dangerous. Both df_p and Df_p are linear maps of $T_p M$ to a 1-dimensional vector space, $\mathbf{R}$ in the first case, $T_{f(p)}\mathbf{R}$ in the second. Moreover, if we identify $T_{f(p)}\mathbf{R}$ with $\mathbf{R}$, as in 4.5, then df and Df coincide, *cf.* 4.5.1.)

If M is endowed with an r-metric and $L: TM \to T^*M$ is the smooth isomorphism from 4.4, then $L^{-1}(df)$ is a vector field. It is denoted ∇f and called the gradient of f. For every vector Y on M we have

(5.5) $\langle \nabla f, Y \rangle = df(Y) = Yf.$

Observe that if $M = \mathbf{R}^m$, then 5.5 is the formula for the derivative in the direction of Y. We are not very far from Calculus yet.

6   Differential Equations on a Smooth Manifold

In the sequel we will often use vector fields to construct various maps. These constructions will be based on the existence theorem for the solutions of systems of ordinary differential equations restated in the context of differentiable manifolds.

In what follows X will be a fixed vector field on M and ∂t will be the vector field on $M \times \mathbf{R}$ consisting of vectors tangent to curves (p_0, t). If ε is a positive function on M, then we set $W_\varepsilon = \{(p, t) \in M \times \mathbf{R} \,||t| < \varepsilon(p)\}$.

(6.1) Theorem *If M is closed, then there is a continuous positive function ε on M and a unique map $f: W_\varepsilon \to M$ such that*

(a) $$f(p, 0) = p,$$

(b) $$Df_{(p,t)}\partial t = X_{f(p,t)}.$$

Uniqueness means that if f_1, δ, W_δ is another set of data satisfying (a) and (b) then $f = f_1$ in $W_\varepsilon \cap W_\delta$.

If M has a boundary, then 6.1 is still valid but with some modifications. This is discussed at the end of this section.

Proof We will interpret the conditions (a), (b) in terms of local coordinates in a single chart U.

Let $\{U_\alpha, h_\alpha\}$ be an adequate atlas on M, let $V_\alpha = h_\alpha^{-1}(D^m)$. Fix a pair U, V and a positive number ε. Then any map $f: V \times [-\varepsilon, \varepsilon] \to U$ is given in local coordinates by m functions $f_1(x, t), \ldots, f_m(x, t), x \in D^m, |t| \leq \varepsilon$. Condition (a) becomes

(a') $$f_i(x, 0) = x, \qquad i = 1, \ldots, m.$$

The vector field ∂t on $V \times [-\varepsilon, \varepsilon]$ has coordinates $(0, \ldots, 0, 1)$. Let the coordinates of X be $a_1(x), \ldots, a_m(x)$. In the same coordinates, the Jacobian of f is given by

$$\begin{pmatrix} \dfrac{\partial f_1}{\partial x_1} & \cdots & \dfrac{\partial f_1}{\partial x_m} & \dfrac{\partial f_1}{\partial t} \\ & \cdots & & \\ \dfrac{\partial f_m}{\partial x_1} & \cdots & \dfrac{\partial f_m}{\partial x_m} & \dfrac{\partial f_m}{\partial t} \end{pmatrix};$$

thus, by 4.2.1, (b) means simply that

(b') $$\frac{\partial f_i}{\partial t}(x, t) = a_i(f_1(x, t), \ldots, f_m(x, t)), \qquad i = 1, \ldots, m.$$

Now, the classical existence theorem [Hu,2.5] asserts that given m smooth functions $a_1, \ldots, a_m$ in $\mathbf{R}^m$, there is a positive number ε and a unique smooth map $f: D^m \times [-\varepsilon, \varepsilon] \to \mathbf{R}^m$, $f = (f_1, \ldots, f_m)$, satisfying (a') and (b'). This means that for every α we have a positive number ε_α and a map $f_\alpha: V_\alpha \times [-\varepsilon_\alpha, \varepsilon_\alpha] \to U_\alpha$ satisfying (a) and (b). Because of the uniqueness, $f_\alpha = f_\beta$ wherever both are defined. Therefore, the theorem follows if we set $\varepsilon = \sum_\alpha \lambda_\alpha \varepsilon_\alpha$, where λ_α is the associated partition of unity. $\square$

We will sometimes call a vector field on M a differential equation and the function f in 6.1 its solution. The curve $f(p, s)$ is called the solution emanating from p. By uniqueness [Hu,l.c], it is completely characterized by the condition $f(p, 0) = p$ and the requirement that its tangent vectors are vectors from X. In particular, if $f(p, s)$ is defined, then $f(f(p, s), t)$ and $f(p, s + t)$ are both solutions emanating from $f(p, s)$. Hence

(6.1.1) $f(f(p, s), t) = f(p, s + t),$ both curves defined in the same range of the parameter t.

This implies that if all solutions are defined for $|t| < \varepsilon_0$, where ε_0 is a positive constant, then they are defined for all values of t. Moreover, in this case the map $f_t: M \to M$ given by $f_t(p) = f(p, t)$ is well-defined for all p and t and 6.1.1 translates to $f_s f_t = f_{s+t}$. Thus f_t is a diffeomorphism with the inverse f_{-t}. All this is certainly the case if M is compact.

(6.2) Corollary   *If M is closed and compact, then a differential equation on it admits a solution $f: M \times \mathbf{R} \to M$. Each map f_t is a diffeomorphism and $f_s f_t = f_{s+t}$ for all s, t.* $\square$

The map $t \mapsto f_t$ is a homomorphism of the additive group of reals into the group of diffeomorphisms of M; such a homomorphism is called a 1-parameter group of diffeomorphisms.

Assume now that M has a non-empty boundary. The proof of 6.1 would fail in that the solutions of the system (b') for $x \in \mathbf{R}^{m-1}$ need not lie in $\mathbf{R}_+$. This can be remedied: Observe first that if X is at the boundary of M and points inside, then its last coordinate a_m is positive. But then $f_m(x, t)$ satisfying (a') and (b') is positive for $t \geq 0$ and small. Similarly, if X points outside then the corresponding solution is in M for $t \leq 0$. Thus, we have the following result:

(6.3) Addendum   *If $\partial M \neq \emptyset$ and X is never tangent to it, then 6.1 remains valid with the map f defined on the set $\{(p, t) \in M \times \mathbf{R} \,|\, -\delta(p) \leq t \leq \varepsilon(p)\},$*

where the functions ε and δ satisfy the following conditions:

(a) $\varepsilon, \delta > 0$ *in* $\operatorname{Int} M$;

(b) *If* $p \in \partial M$ *and* X_p *points inside* M, *then* $\varepsilon(p) > 0$ *and* $\delta(p) = 0$;

(c) *If* $p \in \partial M$ *and* X_p *points outside* M, *then* $\varepsilon(p) = 0$ *and* $\delta(p) > 0$.

Finally, 6.1.1 remains valid and is used in the following exercise:

Exercise Show that if M is compact, $p \in M$ given, then the set of all t for which a solution $f(p, t)$ is defined is a closed subset of $\mathbf{R}$.

7 Collars

As the first application of 6.1 we will prove that ∂M has a neighborhood in M diffeomorphic to $\partial M \times [0, 1)$ under a diffeomorphism identifying ∂M with $\partial M \times \{0\}$. Such a neighborhood is called a collar of ∂M. To prove its existence we note first that

(7.1) there is a vector field X on M which along ∂M points inside M.

For if $\{U_\alpha, h_\alpha\}$ is an adequate atlas, $\{\lambda_\alpha\}$ an associated partition of unity, and ∂_m^α the last coordinate vector rel. the chart U_α, then $X = \sum_\alpha \lambda_\alpha \partial_m^\alpha$ is such a field.

Now, by 6.3 there is a solution $f(p, t)$ of the differential equation X defined for $0 \le t < \varepsilon(p)$ where ε is a continuous positive function on M. Let $W_\varepsilon = \{(p, t) \in M \times \mathbf{R}_+ \,|\, 0 \le t < \varepsilon(p)\}$; f is thus a smooth map of W_ε to M satisfying $f(p, 0) = p$. Since X points inside along ∂M, the differential of f is of maximal rank on $\partial M \times \{0\}$. Therefore (by the Implicit Function Theorem, A,1.1), f is a local diffeomorphism in a neighborhood U of ∂M in W_ε. We now apply the following lemma.

(7.2) Lemma *If* $f \colon U \to V$ *is a local homeomorphism of paracompact spaces which is a homeomorphism on a closed subspace* $C \subset U$, *then* f *is a homeomorphism on a neighborhood of* C. (*For a proof, see* [L, p. 97].) □

Applying this to our situation we deduce the existence of a neighborhood $U_1 \subset U$ on which f is a smooth homeomorphism. But if $g \colon f(U_1) \to U_1$ is the inverse of f then g is smooth: smoothness is a local property and the smoothness of the local inverse of f is assured by the Implicit Function Theorem.

There remains to check that U_1 contains a neighborhood diffeomorphic to $\partial M \times [0, 1)$. But U_1 certainly contains a neighborhood of the form W_γ for some smooth positive function γ. The map $(p, t) \mapsto (p, t/\gamma(p))$ is then a diffeomorphism $W_\gamma \to M \times [0, 1)$.

Observe now that every open neighborhood of ∂M in M is a manifold itself. Thus, we have proved:

(7.3) Theorem *Every open neighborhood of ∂M in M contains a collar neighborhood.* □

(7.4) Corollary *Suppose that ∂M is compact and $\partial M = V_0 \cup V_1$ where V_0, V_1 are closed and disjoint subsets of M. Then there is a smooth function $f: M \to [0, 1]$ such that $f^{-1}(i) = V_i$, $i = 0, 1$.*

Proof Let $W = \partial M \times [0, 1)$ be a collar of ∂M. Define $g: M \to [0, 1]$ by setting

$$g(p) = \begin{cases} t & \text{if } p = (x, t) \in V_0 \times [0, 1/2], \\ 1 - t & \text{if } p = (x, t) \in V_1 \times [0, 1/2], \\ 1/2 & \text{elsewhere in } M. \end{cases}$$

It follows from the compactness of ∂M that g is continuous. Now, a smooth $1/4$-approximation to g, agreeing with g on $\partial M \times [0, 1/2]$, is as desired (*cf.* 2.5). □

(If ∂M is not compact, then the function g in the preceding proof need not be continuous. For instance, let M be the upper half-plane with the origin removed. The boundary of M consists of two rays: $V_0: x_1 > 0$, and $V_1: x_1 < 0$. If the collar is given by the map $(x, t) \mapsto (x, t)$ then g is not continuous. If it is given by $(x, t) \mapsto (x, t |x|)$, g is continuous.)

Observe that the differential of the function f constructed in the proof of 7.4 does not vanish in a neighborhood of ∂M. The requirement that Df vanishes nowhere turns out to be very restrictive.

(7.5) Theorem *Suppose that M is compact, $\partial M = V_0 \cup V_1$, and $f: M \to \mathbf{R}$ is a smooth function such that $f^{-1}(i) = V_i$, $i = 0, 1$. If the differential of f does not vanish, then M is diffeomorphic to $V_0 \times I$.*

Proof Since f cannot have extremal values in the interior of M we must have $f(M) = [0, 1]$. Now, assume that M is a Riemannian manifold and consider the vector field $X = \nabla f / \langle \nabla f, \nabla f \rangle$. It is easy to see that ∇f, hence X, is never tangent to ∂M. Let $g(p, t)$ be a solution of X. Then, $(d/dt)fg = Xf = 1$; hence, $fg(p, t) = t + \alpha(p)$. Setting $t = 0$ we see that $\alpha(p) = f(p)$, i.e.,

$$(*) \qquad\qquad fg(p, t) = t + f(p).$$

Now, consider the set of all $t \in \mathbf{R}$ for which $g(p, t)$ is defined. Since M is compact, it is closed. By $(*)$ it is bounded. Since $g(p, t)$ cannot "stop" at an interior point of M, $(*)$ implies that

$$(**) \qquad g(p, t) \text{ is defined for } -f(p) \le t \le 1 - f(p).$$

In particular, the map $G: V_0 \times I \to M$ given by $G(p, t) = g(p, t)$ is well-defined and smooth. It is a diffeomorphism: Its inverse is given by $p \mapsto (g(p, -f(p)), f(p))$. $\square$

II

Immersions, Imbeddings, Submanifolds

Manifolds appeared in mathematics as submanifolds of Euclidean spaces. In Chapter I we have defined manifolds in an intrinsic way; here, in Section 2, we discuss various ways to define submanifolds and in Section 3 we show that, indeed, every manifold can be realized as a submanifold of a Euclidean space. In Section 4 we introduce isotopy as an equivalence relation for imbeddings and show, in Section 5, that for compact submanifolds isotopy implies a stronger relation, ambient isotopy. In the last section we review briefly the historical development of the notion of a differentiable manifold.

1 Local Equivalence of Maps

The simplest smooth maps are the linear ones and the guiding idea of the differential calculus is to derive information about an arbitrary smooth map from the behavior of a linear map, its differential. We will show that in some circumstances a smooth map is, in a sense, equivalent to its differential. The equivalence in question is local and is defined as follows.

(1.1) Definition Let $f: M \to N$, $g: M_1 \to N_1$, $p \in M$, $q \in M_1$. We say that f at p is equivalent to g at q if there are neighborhoods U of p, V of $f(p)$ and diffeomorphisms h (resp. h_1) of U (resp. V) onto a neighborhood of q (resp. $g(q)$) such that $h_1 f = gh$.

As an example, observe that a linear map $L: \mathbf{R}^m \to \mathbf{R}^n$ of rank k is equivalent to a composition of a projection $(x_1, \ldots, x_m) \mapsto (x_1, \ldots, x_k)$ with the inclusion $(x_1, \ldots, x_k) \mapsto (x_1, \ldots, x_k, 0, \ldots, 0)$.

As another example, let $f: \mathrm{Gl}(n) \to \mathrm{Gl}(n)$ be the map sending a matrix A to $A^t A$. For any two matrices A, B, f at A is equivalent to f at B. For if $h: \mathrm{Gl}(n) \to \mathrm{Gl}(n)$ is the multiplication on the left by BA^{-1} and $h_1: \mathrm{Gl}(n) \to \mathrm{Gl}(n)$ is the multiplication on the left by BA^{-1} and on the right by $^t(BA^{-1})$, then $h(A) = B$ and $h_1 f = fh$.

Exercise Let $S(n)$ be the set of symmetric matrices viewed as an open subset of $\mathbf{R}^{n(n+1)/2}$, and let $f: S(n) \to S(n)$ be given by $f(M) = M^k$, k a a positive integer. Show that for every matrix $M \in S(n)$ there is a diagonal matrix D such that f at M is equivalent to f at D.

Now, let $f: M \to N$ be a smooth map; the dimension of $Df(T_p M)$ is called the *rank* of f at p.

(1.2) Proposition *If the rank of f is constant in a neighborhood of p, then f at p is equivalent to Df_p at $\mathbf{0}$. In particular, if f is of maximal rank at $p \in M$, then f is locally equivalent at p to either a standard projection or a standard inclusion.*

Proof By I,4.2.1 the rank of f equals the rank of the Jacobian of f with respect to some charts about p and $f(p)$. Thus the proposition follows from A,1.2 and A,1.3. □

2 Submanifolds

Intuitively, a submanifold M of a manifold N is a subset of N which is a manifold and which locally looks like $\mathbf{R}^m$ in $\mathbf{R}^n$, at least if both M and N are closed. We will show in 2.3 that this intuition justifies the following batch of definitions.

(2.1) *Definition* Let $f: M \to N$ be a smooth map. We say that f is an *immersion* if Df is everywhere injective, a *submersion* if Df is everywhere surjective. We say that f is an *imbedding* if f is an immersion and a topological imbedding. $M \subset N$ is a *submanifold* if the inclusion map is an imbedding.

Clearly, if $f: M \to N$ is an imbedding then $f(M)$ (with the differentiable structure induced by f) is a submanifold. This is not in general true if f is only a one-to-one immersion. An example of some interest is as follows: Let α be a real number and f_α the imbedding of $\mathbf{R}$ in $\mathbf{R}^2$ as the line $y = \alpha x$; let $\pi: \mathbf{R}^2 \to S^1 \times S^1$ be the covering map $(x, y) \mapsto (\exp(2\pi i x), \exp(2\pi i y))$. Then, the composition πf_α is an immersion, which is one-to-one if α is irrational. But the image is then a dense subset of the torus and—with the topology of a subset—is not even a topological manifold.

Exercise Suppose that M is a smooth manifold and a closed subset of a smooth manifold N. Show that M is a submanifold of N if and only if the following holds: A function f on M is smooth if and only if it is a restriction to M of a smooth function on N.

We intend now to characterize these subsets of a manifold N which can be given the structure of a submanifold. In the case of manifolds with boundary we will consider only submanifolds imbedded in a particularly nice way.

It will be convenient here to write $\bar{\mathbf{R}}^m$ for the subspace of the last m coordinates in $\mathbf{R}^n$; let $\bar{\mathbf{R}}_+^m = \mathbf{R}_+^n \cap \bar{\mathbf{R}}^m = \{(x_1, \ldots, x_n) \in \mathbf{R}^n \mid x_1, \ldots, x_{n-m} = 0, x_n \geq 0\}$.

(2.2) *Definition* A submanifold $M \subset N$ is *neat* if it is a closed subset of N and:

(a) $M \cap \partial N = \partial M$;
(b) At every point $p \in \partial M$ there is a chart (U, h) in N, $h: U \to \mathbf{R}_+^n$, such that $h^{-1}\bar{\mathbf{R}}_+^m = U \cap M$.

Condition (b) means, intuitively, that ∂M meets ∂N like $\bar{\mathbf{R}}_+^m$ meets $\mathbf{R}^{n-1}$. Observe also that the only neat submanifolds of a closed manifold are closed manifolds imbedded as closed subsets.

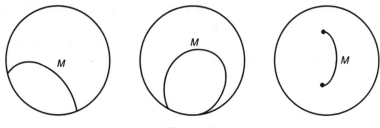

Figure II,1

(2.3) Theorem *A closed subset $M \subset N$ can be given a structure of a neat m-dimensional submanifold if and only if at every point p of M there is a chart (U, h) in N satisfying either of the following two conditions:*

(a) *If $p \in \text{Int } N$, then $h^{-1}\bar{\mathbf{R}}^m = U \cap M$; if $p \in \partial N \cap M$, then $h^{-1}\bar{\mathbf{R}}^m_+ = U \cap M$.*

(b) *there is a submersion $\sigma: U \to \mathbf{R}^{n-m}$, which is also a submersion on $U \cap \partial N$, such that $\sigma^{-1}(0) = U \cap M$.*

This structure on M is unique up to a diffeomorphism.

Proof If $M \subset N$ is a neat submanifold, then by 1.2 the inclusion map is locally equivalent to the standard inclusion $\mathbf{R}^m \hookrightarrow \mathbf{R}^n$; this yields at every interior point of M a chart satisfying condition (a); the existence of such a chart at the boundary points is a part of the definition.

Assume now that (a) holds at a point $p \in \text{Int } M$ and let $\pi: \mathbf{R}^n \to \mathbf{R}^{n-m}$ be the standard projection of $\mathbf{R}^n$ onto $\mathbf{R}^{n-m}$. Then πh is a submersion and

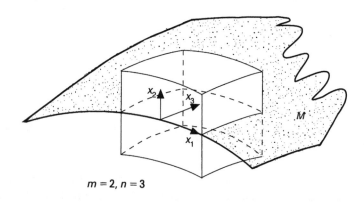

$m = 2, n = 3$

Figure II,2

$h^{-1}\pi^{-1}(0) = h^{-1}(\bar{\mathbf{R}}^m) = U \cap M$. If $p \in \partial N$, then h maps $U \cap \partial N$ diffeomorphically into $\mathbf{R}^{n-1}$; hence πh is a submersion on $U \cap \partial N$, i.e., (b) holds at p.

Suppose that (b) holds at p and let $p \in M \cap \partial N$. Then h sends $U \cap \partial N$ to $\mathbf{R}^{n-1}$; hence σh^{-1} and $\sigma h^{-1}|\mathbf{R}^{n-1}$ are both submersions. By 1.2 there is a diffeomorphism $f: (\mathbf{R}^n, \mathbf{R}^n_+) \to (\mathbf{R}^n, \mathbf{R}^n_+)$ such that $\sigma h^{-1}f$ is a standard projection. Let $g = f^{-1}h$. Then $g(U) \subset \mathbf{R}^n_+$, $\sigma g^{-1}(\bar{\mathbf{R}}^m) = \mathbf{0}$ and $g\sigma^{-1}(\mathbf{0}) \subset \bar{\mathbf{R}}^m$. Thus (U, g) is a chart and, since $g\sigma^{-1}(\mathbf{0}) = g(U \cap M) \subset \mathbf{R}^n_+ \cap \bar{\mathbf{R}}^m = \bar{\mathbf{R}}^m_+$, $\sigma^{-1}(\mathbf{0}) \subset h^{-1}(\bar{\mathbf{R}}^m)$. The reverse inclusion is obvious; hence $U \cap M = h^{-1}(\bar{\mathbf{R}}^m_+)$, i.e., (a) holds. If $p \in \text{Int } M$, then the proof is similar but simpler.

Assume that (a) is satisfied at every point of M. We will show that M can be given a structure of a manifold. Cover M by charts (U_α, h_α) satisfying (a) and set $V_\alpha = U_\alpha \cap M$, $g_\alpha = h_\alpha | V_\alpha$; we claim that $\{V_\alpha, g_\alpha\}$ yields an atlas on M.

Suppose that $V_\alpha \cap V_\beta = U_\alpha \cap U_\beta \cap \text{Int } M \neq \emptyset$. Then $h_\alpha h_\beta^{-1}$: $h_\beta(U_\alpha \cap U_\beta) \to \mathbf{R}^n$ is a smooth map and $g_\alpha g_\beta^{-1}$ is its restriction to $W = h_\beta(V_\alpha \cap V_\beta) \subset \bar{\mathbf{R}}^m$. Since $h_\beta^{-1}(W) \subset M$, $g_\alpha g_\beta^{-1}(W) = h_\alpha h_\beta^{-1}(W) \subset \bar{\mathbf{R}}^m$, i.e., $g_\alpha g_\beta^{-1}$ is a smooth map of an open subset of $\bar{\mathbf{R}}^m$ into $\bar{\mathbf{R}}^m$. If $p \in \partial M$, then the same argument works with $\bar{\mathbf{R}}^m$ replaced by $\bar{\mathbf{R}}^m_+$. Thus M is a differentiable manifold.

There remains to show that it is a submanifold, i.e., that the inclusion $M \subset N$, which is certainly a topological imbedding, is an immersion. But with V_α, U_α as defined, the inclusion $V_\alpha \hookrightarrow U_\alpha$ is locally equivalent either to the inclusion $\mathbf{R}^m \subset \mathbf{R}^n$ or $\bar{\mathbf{R}}^m_+ \subset \mathbf{R}^n$.

To prove the uniqueness, let $\mathcal{U} = \{U_\alpha, h_\alpha\}$, $\mathcal{V} = \{V_\beta, g_\beta\}$ be two smooth structures on M, both giving it the structure of a submanifold. Suppose that $\emptyset \neq U_\alpha \cap V_\beta \subset U \cap \text{Int } M$, where (U, h) is a chart in N as in (a). Since M is a submanifold, hg_β^{-1} is a smooth map of maximal rank. By (a), $hg_\beta^{-1}(g_\beta(V_\beta)) \subset h(U \cap M) \subset \bar{\mathbf{R}}^m$. Thus hg_β^{-1} is a smooth map of an open subset of $\bar{\mathbf{R}}^m$ into $\bar{\mathbf{R}}^m$ of rank m. By the Implicit Function Theorem (A,1.1) its inverse, $g_\beta h^{-1}|\bar{\mathbf{R}}^m$, is smooth too. This implies that $g_\beta h_\alpha^{-1} = g_\beta h^{-1}hh_\alpha^{-1}$ is smooth, i.e., the atlases $\mathcal{U}$, $\mathcal{V}$ are compatible. A similar argument works at the boundary points. $\square$

We apply 2.3 to study inverse images of points under a smooth map.

(2.4) Definition Let $f: M \to N$ be a smooth map. We say that $q \in N$ is a *regular value* of f if Df is surjective at every point $p \in f^{-1}(q)$ and $Df|T(\partial M)$ is surjective at every point $p \in f^{-1}(q) \cap \partial M$.

(2.5) Corollary   *If q is a regular value of f, then $f^{-1}(q)$ is a neat submanifold of M.*

Proof   Note that if Df is surjective at p, then it is surjective—hence f is a submersion—in a neighborhood of p. Hence the corollary follows from 2.3(b).   □

As an example, consider a map $f: \mathbf{R}^n \to \mathbf{R}$ given by $(x_1, \ldots, x_n) \mapsto \sum_i x_i^2$. Then every $r \neq 0$ is a regular value; in particular, we get a submanifold structure on $S^{n-1} = f^{-1}(1)$. This is still the standard structure: The construction in I,1.2 did, in fact, represent S^{n-1} as a submanifold (just consider $(h^{\pm})^{-1}$); thus the uniqueness part of 2.3 applies.

We can generalize this procedure to the tangent bundle of a manifold M, endowed with a Riemannian metric $\langle\ ,\ \rangle$, and the function $f: TM \to \mathbf{R}, f(v) = \langle v, v \rangle$. If $v = \sum_i v_i\, \partial_i$ in local coordinates, then $f(v) = \sum_{i,j} a_{ij} v_i v_j$ for some symmetric positive definite matrix (a_{ij}). Thus every $r \neq 0$ is a regular value of f. The manifold $f^{-1}(1)$ is called the tangent unit sphere bundle of M.

The following corollary is more general than 2.5. We assume for simplicity that $\partial M = \emptyset$.

(2.6) Corollary   *Let $f: M \to N$ be smooth and assume that, for some $q \in N$, f is of constant rank on a neighborhood of $f^{-1}(q)$. Then $f^{-1}(q)$ is a submanifold of M.*

Proof   Let $p \in f^{-1}(q)$ and assume that the rank of f is k. By 1.2 there is a chart (U, h) about p in M and a chart (V, g) about q in N such that gfh^{-1} is the standard projection $\mathbf{R}^m \to \mathbf{R}^k$ followed by the inclusion $\mathbf{R}^k \hookrightarrow \mathbf{R}^n$. In particular, gf seen as a map into $\mathbf{R}^k$ is a submersion and the corollary follows from 2.3(b).   □

As an example, consider the map $f: \mathbf{Gl}(n) \to \mathbf{Gl}(n)$ sending the matrix A to $A^t A$. We have shown in 1.2 that for any two matrices A, B, f at A is equivalent to f at B. Thus f is of constant rank on $\mathbf{Gl}(n)$. Since $f^{-1}(I_n) = \mathbf{O}(n)$, we obtain a smooth structure on $\mathbf{O}(n)$ as a submanifold of $\mathbf{Gl}(n)$. Now, the multiplication in $\mathbf{O}(n)$ is the restriction to $\mathbf{O}(n)$ of the multiplication in $\mathbf{Gl}(n)$. We have already noted that the latter is smooth. Since $\mathbf{O}(n)$ is a smooth submanifold and the restriction of a smooth map to a smooth submanifold is smooth, it is a Lie group (*cf.* I,1.7).

Exercise Let $F: \mathrm{Gl}(n) \to \mathbf{R}^{n(n+1)/2}$ be given by $F(M) = (f_{11}, \ldots, f_{1n}, f_{22}, \ldots, f_{2n}, \ldots, f_{nn})$ where for a given matrix M the matrix $(f_{ij}) = M^t M$. Show that $F(I_n)$ is a regular value of F.

Exercise Show that the set of $m \times n$ matrices with orthonormal rows is a submanifold of $\mathbf{R}^{mn}$.

(2.7) As another example of the application of 2.5 consider a map $f: (M, \partial M) \to (N, \partial N)$ where M and N are both compact, connected, and of the same dimension n. If $q \in \mathrm{Int}\, N$ is a regular value of f then $f^{-1}(q)$ is a compact 0-dimensional submanifold of $\mathrm{Int}\, M$, that is, a finite set of points $p_1, p_2, \ldots, p_k$, and the map f is a local diffeomorphism in a neighborhood of each p_i. Suppose that M and N are oriented, and let γ_M, γ_N be corresponding generators of $H_n(M, \partial M)$, $H_n(N, \partial N)$. Then

$$f_* \gamma_M = d\gamma_N,$$

and the integer d is called the degree of f.

The degree of f can be calculated from the behavior of f at points p_i. For if γ_i is the induced local orientation at p_i, $i = 1, 2, \ldots, k$, and γ is the induced local orientation at q, then $f_* \gamma_i = d_i \gamma$, where $d_i = \pm 1$. An elementary argument, left as an exercise, shows that

(2.7.1) $$d = \sum_i d_i.$$

The interest and the importance of this stems from the Brown–Sard Theorem (A,3.1): The regular values always exist.

(2.8) If $M \subset N$ is a neat submanifold, $p \in \partial M$, and U is a chart as in 2.2(b), then $T_p M \subset T_p N$ is represented by the subspace of the last m coordinates and $T_p \partial N$ by the space of the first $n - 1$ coordinates. Thus

(2.8.1) $\qquad\qquad T_p M$ and $T_p \partial N$ are in general position,

which condition is independent of the choice of the chart U.

In fact, this condition is equivalent to 2.2(b). To see this we prove first that

(2.8.2) if $M \subset N$ is a neat submanifold, then there is a collar of ∂N in N such that its restriction to ∂M is a collar of ∂M in M.

This is proved in the same way as I,7.3 except that to construct the vector field X in I,7.1 pointing inside N one first constructs on M a vector field pointing inside M along the boundary. By 2.8.1 it points inside N as well; thus it can be extended over N to yield X.

Now let $p \in \partial M$. There is then a diffeomorphism g of $(\mathbf{R}^{n-1}, \bar{\mathbf{R}}^{m-1})$ onto a neighborhood V of p in ∂N such that $g(\bar{\mathbf{R}}^{m-1}) = V \cap \partial M$. By 2.8.2 there is a diffeomorphism f of $V \times \mathbf{R}_+$ onto a neighborhood U of p in M such that $f((V \cap \partial M) \times \mathbf{R}_+) = U \cap M$. Let $h(x, t)$, $x \in \mathbf{R}^{n-1}$, $t \in \mathbf{R}_+$, be given by $h(x, t) = f(g(x), t)$. If $x \in \bar{\mathbf{R}}_+^m$, then $x = (x', t)$, $x' \in \bar{\mathbf{R}}^{m-1} \subset \mathbf{R}^{n-1}$, $t \in \mathbf{R}_+$, thus $g(x') \in V \cap \partial M$, and $h(x', t) = f(g(x'), t) \in U \cap M$. This shows that p has neighborhoods satisfying 2.2(b). $\square$

3   Imbeddings in $\mathbf{R}^n$

Most manifolds we have considered up to now were submanifolds of a Euclidean space. The question arises whether this is true in general, that is, whether every smooth manifold can be imbedded in some Euclidean space. This was answered affirmatively by H. Whitney in [Wi2]. We prove here only a weak form of Whitney's theorem.

(3.1)   Theorem   *A compact smooth manifold can be imbedded in a Euclidean space.*

Proof   If M is compact, it has a finite adequate atlas $\{U_i, h_i\}$, $i = 1, \ldots, k$. Let $\mu_1, \ldots, \mu_k$ be an associated family of bump functions (I,2.2). Define maps $\phi_i: M \to \mathbf{R}^m$ by $\phi_i = \mu_i h_i$ and let $h: M \to \mathbf{R}^n$, $n = k(m + 1)$, be given by

$$h(p) = (\phi_1(p), \ldots, \phi_k(p), \mu_1(p), \ldots, \mu_k(p))$$

$$\in \mathbf{R}^m \times \cdots \times \mathbf{R}^m \times \mathbf{R} \times \cdots \times \mathbf{R}.$$

We will show that h is an imbedding. Let $p \in M$ and let i be so chosen that $\mu_i(p) = 1$. If $h(p) = h(q)$, then also $\mu_i(q) = 1$; hence $h_i(p) = h_i(q)$. Since the h_i are homeomorphisms this implies that $p = q$. Since M is compact, h is a topological imbedding. To see that it is of maximal rank, note that the Jacobian matrix at p in the coordinate system U_i contains a unit $m \times m$ matrix. $\square$

The dimension of the imbedding space is absurdly high: For the projective space P^m, with the differentiable structure given by m charts as in I,1.3, this theorem yields an imbedding in $\mathbf{R}^{m(m+1)}$. In fact, Whitney's theorem [Wi2, Theorem 5] asserts the following:

(3.2) Theorem *Let* $f: M \to N$ *be a smooth map which is an imbedding on a closed subset* $C \subset M$; *let* ε *be a continuous positive function on* M. *If* $\dim M \geq 2 \dim N + 1$, *then there is an imbedding* $g: M \to N$ ε-*approximating* f *and such that* $f \mid C = g \mid C$.

The results of Whitney were greatly generalized by A. Haefliger, *cf.* [H1].

4 Isotopies

We introduce now a notion of equivalence for imbeddings. Naturally enough, we want two imbeddings to be equivalent if one can be deformed to the other through imbeddings. The most convenient form of stating this precisely is as follows.

(4.1) *Definition* Let $f, g: M \to N$ be two imbeddings. An *isotopy* between f and g is a smooth map $F: M \times \mathbf{R} \to N$ such that

  (a) $F(x, 0) = f(x), F(x, 1) = g(x)$;
  (b) $F_t = F \mid M \times (t)$ is an imbedding for $0 \leq t \leq 1$.

The assumption that $F(x, t)$ is defined for all t is for technical convenience; its actual behavior for $t < 0$ and $t > 1$ is of no importance: We can always assume that $F(x, t) = f(x)$ for $t \leq 0$ and $= g(x)$ for $t \geq 1$. For let $\mu(t)$ be a smooth non-decreasing function such that $\mu(t) = 0$ for $t \leq 0$ and $= 1$ for $t \geq 1$. Then $F(x, \mu(t))$ is an isotopy that is constantly $f(x)$ for $t \leq 1$ and $g(x)$ for $t \geq 1$.

Isotopy is an equivalence relation: To see that it is transitive suppose that F is an isotopy between f and g and that G is an isotopy between g and h. Let μ_1, μ_2 be diffeomorphisms of $\mathbf{R}$ onto itself such that μ_1 maps the segment $[0, 1/3]$ onto $[0, 1]$ and μ_2 does the same to the segment $[2/3, 1]$. Assume also that F is constantly g for $t \geq 1$ and G is constantly g for $t \leq 0$. Then an isotopy between f and h is given by

$$H(x, t) = \begin{cases} F(x, \mu_1(t)) & \text{if } t \leq 1/2; \\ G(x, \mu_2(t)) & \text{if } t \geq 1/2. \end{cases}$$

Every isotopy $F(x, t)$ induces a level preserving imbedding $M \times \mathbf{R} \rightarrow N \times \mathbf{R}$. The converse is also true:

(4.2)   Lemma   *Let $G: M \times \mathbf{R} \rightarrow N \times \mathbf{R}$ be a level preserving imbedding, i.e., $G(x, t) = (F(x, t), t)$. Then F is an isotopy.*

Proof   Let U_α be a chart in M and V_β be a chart in N. Then the Jacobian of G with respect to charts $U_\alpha \times \mathbf{R}$, $V_\beta \times \mathbf{R}$ is the matrix

$$\begin{pmatrix} J_\beta^\alpha F_t & * \\ 0, \dots, 0 & 1 \end{pmatrix},$$

where $F_t = F \,|\, M \times (t)$ and $J_\beta^\alpha F_t$ is its Jacobian with respect to U_α, V_β. Since the Jacobian of G is of rank $m + 1$, the rank of $J_\beta^\alpha F_t$ must be m. The inverse of F_t on the image is constructed by lifting it to the level t and applying G^{-1}.   □

Assume now M to be a closed manifold. By I,6.2 a vector field on M admitting a global solution induces an isotopy of the identity map of M: its solution. This connection between vector fields and isotopies is best described in terms of vector fields on $M \times \mathbf{R}$. First, define the t-coordinate of a vector on $M \times \mathbf{R}$ to be its image under the differential of the projection $\pi: M \times \mathbf{R} \rightarrow \mathbf{R}$. Certainly, if $G: M \times \mathbf{R} \rightarrow M \times \mathbf{R}$ is a level preserving imbedding, then $DG(\partial t)$ is a vector on $M \times \mathbf{R}$ with the t-coordinate ∂t. The next lemma asserts the converse.

(4.3)   Lemma   *If X is a vector field on $M \times \mathbf{R}$ with the t-coordinate ∂t and admitting a global solution, then X induces an isotopy of the identity map of M.*

Proof   Let $H: (M \times \mathbf{R}) \times \mathbf{R} \rightarrow M \times \mathbf{R}$ be the global solution and let $G(p, t) = H(p, 0, t)$. Since H is a solution, $H((p, x), 0) = (p, x)$, which implies

(∗) $G(p, 0) = (p, 0).$

Now, we will show that

(∗∗) G is level preserving.

Write $H = (H_1, H_2)$ where H_2 is a composition of H with the projection π of $M \times \mathbf{R}$ on $\mathbf{R}$. Thus $DH_2(\partial t) = D\pi(X) = \partial t$, which implies $\partial H_2/\partial t =$

$dt/dt = 1$. Consequently $H_2((p, x), t) = t + \beta(p, x)$ and

$$G(p, t) = (H_1((p, 0), t), t + \beta(p, 0)).$$

Comparing this with $(*)$ we see that $\beta(p, 0) = 0$, which proves $(**)$.

Now, consider the map $L: M \times \mathbf{R} \to (M \times \mathbf{R}) \times \mathbf{R}$ that sends the point (p, t) to $(H((p, t), -t), t)$. We have

$$L(G(p, t)) = (H(H(p, 0, t), -t), t) = (H(p, 0, t - t), t) = (p, 0, t).$$

It follows that L (followed by an obvious projection) is an inverse of G, i.e., G is a diffeomorphism. By 4.2 this concludes the proof of the lemma.   □

We will consider a few examples.

(4.4)   Let $M \in \mathbf{Gl}(n)$ and consider the linear diffeomorphism $f_M : \mathbf{R}^n \to \mathbf{R}^n$, given by $f_M(v) = M \cdot v$. Assume that $\det(M) > 0$ and let $M(t)$ be a smooth path in $\mathbf{Gl}(n)$ such that $M(0) = M$, $M(1) = I_n$. Then $F(v, t) = M(t) \cdot v$ defines an isotopy of f_M to the identity map. If $\det(M) < 0$, then an analogous construction yields an isotopy of f_M to the map that reverses the first coordinate and preserves the rest.

(4.5)   Now let $f: \mathbf{R}^n \to \mathbf{R}^n$, $f = (f^1, \ldots, f^n)$, be a diffeomorphism satisfying $f(\mathbf{0}) = \mathbf{0}$. By A,2.1 we can write $f^j(x) = \sum_i a_i^j(x)x_i$. Let $F(x, t) = (1/t)f(tx)$. Since $(1/t)f^j(tx) = \sum_i a_i^j(tx)x_i$, F is a smooth map for all t. For $t \neq 0$ it clearly is a diffeomorphism; to see what it is for $t = 0$ note that $a_i^j(\mathbf{0}) = (\partial f_j/\partial x_i)(\mathbf{0})$. Thus, $F(x, 0) = (Jf) \cdot x$ is a linear map given by the Jacobian of f at $\mathbf{0}$. It follows that f is isotopic to the linear map given by its Jacobian matrix at $\mathbf{0}$. As we noticed in 4.4, this last map is in turn isotopic either to the identity map or to an elementary orientation reversing map.

(4.6)   An example of an isotopy is provided in the proof of Theorem I,3.3: What we have actually shown there is that every isomorphism between two Riemannian bundles is isotopic to an isometry.

Exercise   Show that every orientation preserving diffeomorphism of S^1 is isotopic to the identity map.

An analogous theorem is not true for spheres of dimension ≥ 6.

(4.7) A general theorem concerning isotopies was proved by H. Whitney [Wi2, Theorem 6]:

Theorem *Let $f, g: M^m \to N^n$ be two homotopic imbeddings of a compact manifold M^m. If $n \geq 2m + 2$, then f and g are isotopic.*

5 Ambient Isotopies

Two imbeddings $f, g: M \to N$ might well be isotopic without the complements $N - f(M)$, $N - g(M)$ being homeomorphic. (For instance, the complement in $\mathbf{R}^2$ of a circle with a point removed is not homeomorphic to the complement of the open segment $(0, 1) \subset \mathbf{R}^1 \subset \mathbf{R}^2$.) It is even possible for this to happen when both $f(M)$, $g(M)$ are closed subsets of N (see 5.2). This is the rationale for a stronger notion of equivalence of imbeddings, that of an ambient isotopy.

(5.1) *Definition* Let $f, g: M \to N$ be two imbeddings. An *ambient isotopy* between f and g is an isotopy $F: N \times \mathbf{R} \to N$ such that $F(p, 0) = p$, $F(f, 1) = g$.

We have just seen that two isotopic imbeddings of M need not be ambient isotopic. It is a consequence of our next theorem, the Isotopy Extension Theorem, that they are ambient isotopic if M is compact and N closed. This was proved first by R. Thom [T4]; a stronger version is due to R. Palais [Pa2] and, independently, to J. Cerf [C1].

(5.2) **Theorem** *Let $f: M \to N$ imbed M in a closed manifold N; let $K \subset M$ be a compact subset and $G: M \times \mathbf{R} \to N \times \mathbf{R}$ an isotopy of f. Then there is an isotopy $H: N \times \mathbf{R} \to N \times \mathbf{R}$ of the identity map of N such that $H(f(x), t) = G(x, t)$ for $x \in K, t \in [0, 1]$.*

Proof The construction of G will be based on Lemma 4.3.

First, consider the vector field $X = DG(\partial t)$. It is defined on $G(M \times \mathbf{R})$ and, since $G(M \times \mathbf{R})$ is a smooth submanifold of $N \times \mathbf{R}$, it is locally extendable. Now, $B = G(K \times [0, 1])$ is compact, hence a closed subset of $N \times \mathbf{R}$. By I,3.4 $X | B$ extends over $N \times \mathbf{R}$ to a vector field Y. Clearly, we can assume that Y vanishes outside of some neighborhood U of B with compact closure. Finally, define a vector field Z by $Z = (Y - t$-coordinate

of Y) + ∂t. Clearly

$(*)$ the t-coordinate of Z equals ∂t.

Since $Y|B = X|B$, the t-coordinate of $Y|B$ is ∂t; thus

$(**)$ $Z|B = X|B$.

It remains to be shown that

$(***)$ Z admits a global solution.

Let V be an open set with compact closure containing $\mathrm{Cl}(U)$. Then there is a constant ε such that all solutions originating in V are defined for $|t| < \varepsilon$. On the other hand, solutions originating outside V are—until they reach U—simply curves $(x_0, t + t_0)$. Since $\mathrm{Cl}(U)$ is compact, there is a positive constant δ such that they are all defined at least for $|t| < \delta$. This implies $(***)$, cf. I,6.2.

By $(*)$ and $(***)$, Lemma 4.3 applies and yields an isotopy H of the identity map of N. Consider a solution curve C of X which at $t = 0$ passes through $(f(x_0), 0)$, i.e., the curve $G(x_0, t)$. Because of $(**)$, for $0 \le t \le 1$, C is also a solution of Z, i.e., the curve $H(f(x_0), t)$. Thus $H(f(x), t) = G(x, t)$ for $(x, t) \in K \times [0, 1]$. $\square$

Note that the isotopy H is stationary outside of a compact set.

If $\partial N \ne \emptyset$ but the isotopy G moves K in the interior of N, then the theorem remains valid: Apply 5.2 to the interior of N and note that since H is stationary outside of a compact subset of $\mathrm{Int}\, N$ it can be extended over N by requiring it to be stationary on N.

The assumption that K is compact is essential and cannot be replaced by requiring $f(M)$ to be closed in N and K closed in M.

Exercise Consider an imbedding $\mathbf{R} \to \mathbf{R}^3$ where the image is "the line with a knot":

(a) Show that this imbedding is isotopic to the standard imbedding $\mathbf{R} \subset \mathbf{R}^3$ but not ambient isotopic to it. (*Hint*: a consideration of one point compactification of $\mathbf{R}^3$ shows that the complement of the "knotted line" is homeomorphic to the complement of the trefoil knot.) (b) Trace the proof of 5.2 in this case. What goes wrong?

(5.3) Corollary *If M is connected and p, q ∈ M, then there is an isotopy F_t of the identity map of M such that $F_1(p) = q$.*

Proof By I,2.6 there is a smooth path joining p and q. Since such a path can be considered an isotopy of the inclusion of p in M, the corollary follows from 5.2. □

(5.4) Corollary *If $f: S^m \to S^n$ is an imbedding and $n \geq 2m + 2$, then f extends to an imbedding of D^{m+1}.*

Proof f is homotopic to the standard imbedding; hence by 4.7 it is isotopic to it. Since the isotopy is ambient and the standard imbedding extends to an imbedding of D^m, so also does f.   □

Of course, 5.4 is true with S^n replaced by any manifold with vanishing m-dimensional homotopy group.

6 Historical Remarks

Differentiable manifolds entered mathematics as curves in the plane and surfaces in $\mathbf{R}^3$. Leaving aside the theory of curves, which had its own peculiarities, it is interesting to trace the developments that culminated in the notion of a differentiable manifold.

A salient point is that at the early stage there was no perceived need to *define* a surface. The surface was a geometric object which simply *was* there; the task of a mathematician was to find an analytical way to deal with it, to describe it. Thus Euler in [E1, pp. 324–325], discusses how a surface determines its equation, which will "express its nature." His method suggests that he wants the surface to be locally a graph, but he does allow the possibility that the z coordinate is not uniquely determined by x and y. He also recognizes that the representation is local; the surfaces that consist of patches, each given by a different equation, are called "discontinuas seu irregulares." This 1748 book seems to be the first place where a notion of a general surface appears. Somewhat later, in 1771, Euler gives a definition of a surface using a parametric representation. This would correspond to our notion of an imbedding of $\mathbf{R}^2$ in $\mathbf{R}^3$, but for the fact that no regularity conditions are explicitly stated [E2].

Toward the end of the 18th century G. Monge uses an equation $F(x, y, z) = 0$ to represent a surface (and two such equations to represent a curve). This was again a local representation only, but nothing more general was needed. The research was centered either on local properties of surfaces or on algebraic surfaces, generally quadrics, given, indeed, by one equation.

This point of view did not change until Poincaré. For instance, Gauss, in his fundamental paper of 1832 ([G]), distinguishes two ways of defining a surface: The first way is by an equation $W(x, y, z) = 0$; the second by a system of equations $x = f(p, q)$, $y = g(p, q)$, $z = h(p, q)$. Again, no more than one local coordinate patch is considered and no regularity conditions are explicitly stated though they are implicitly assumed as it is clear from subsequent computations.

Up to the time of B. Riemann manifolds thus appear only as curves or surfaces in $\mathbf{R}^3$. While Riemann is generally credited with the idea of an abstract n-dimensional manifold, what actually appears in [R] is what we would call one chart with a metric given by a linear element ds^2. There is no general definition of a manifold (Riemann seems to consider this a philosophical problem), but an important step has been taken in that his objects are n-dimensional and not necessarily submanifolds of anything else.

The "modern" definition of a differentiable manifold appears for the first time in the 1895 paper of H. Poincaré [P1]. Manifolds are still submanifolds of $\mathbf{R}^n$, but all necessary elements are present: The definition is by overlapping charts and the condition on the rank of the Jacobian is stated explicitly. (Strictly speaking, what he defines we would call a 1-1 immersed submanifold of $\mathbf{R}^n$.)

Poincaré also considers inverse images of regular values ("première definition"), and shows that they are manifolds (i.e., our 2.5), but that not every submanifold can be defined in such a way.

Manifolds and homeomorphisms considered by Poincaré were always smooth. As a matter of fact, he always requested analyticity, but when constructing examples (e.g., of what is now called a Poincaré sphere, [P3]) he never verified that they were smooth manifolds, or submanifolds of Euclidean space. Actually, he did not need smoothness but only triangulation and provided in [P2] an incomplete proof of the possibility of triangulation of a smooth manifold. (Poincaré's pioneering work is discussed in detail in [Di].)

Poincaré also considered manifolds constructed by identifications on faces of 3-dimensional cells and gave a criterion for such an identification to lead to a manifold.

During the next 30 years the concept of a topological and triangulated manifold was formulated with necessary precision, but until the 1930s smooth manifolds were still considered in reference to some imbedding (e.g., H. Hopf [Ho1]). The first intrinsic definition of a manifold—that is, not as a submanifold of anything else—appeared, in a rather awkward form as a set of axioms, in the work of O. Veblen and J. H. C. Whitehead ([VW]) in 1931. In the present form, it can be found in the book of P. Alexandroff and H. Hopf [AH] and in the papers of H. Whitney [Wi2] and J. W. Alexander [A]. In his paper Whitney showed that smooth manifolds can always be imbedded in a Euclidean space, hence that the intrinsic definition was not more general than one given by Poincaré.

The question of relations between smooth and triangulated or topological manifolds was posed by Alexander [A]. After expressing confidence that the triangulation of smooth manifolds is only a question of "honest toil," Alexander asked about triangulability of topological manifolds and about the validity of the so-called Hauptvermutung for manifolds. At about the same time, Alexandroff and Hopf asked whether every triangulated manifold carries a smooth structure. Thus the questions were clearly posed but, with the exception of the "honest toil" performed by S. S. Cairns in 1934 [Ca], the answers were slow in coming.

The first answer came in 1959 when M. Kervaire gave an example of a combinatorial manifold that is not smoothable [K2]. The triangulability of topological manifolds had to wait until 1969 when L. Siebenmann provided a counterexample [Si]. Thus the three classes of manifolds—topological, triangulated, smooth—are all distinct.

There remained the question of the Hauptvermutung. For smooth manifolds J. H. C. Whitehead, improving on the work of Cairns, showed in 1940 that if two smooth manifolds are diffeomorphic, then their C^1 triangulations are indeed combinatorially equivalent [Wh3]. The assumption of diffeomorphism cannot be weakened: In 1969, L. Siebenmann gave examples of smooth 5-dimensional manifolds homeomorphic but with non-equivalent combinatorial triangulations. The examples are not difficult to construct but the proof that they possess desired properties requires deep investigation into the structure of topological manifolds. The appropriate machinery, due to R. Kirby and L. Siebenmann, is described in [KS].

III

Normal Bundle, Tubular Neighborhoods

A curve in $\mathbf{R}^3$ is contained in a nice, tube-like neighborhood. In Section 2 we will show that submanifolds of a smooth manifold always possess similar neighborhoods called, by analogy, tubular neighborhoods.

Tubular neighborhoods are not unique, but any two are related by an isotopy of the entire manifold; this is the content of the Tubular Neighborhood Theorem, proved in Section 3. A special case of it is used in Section 6 to define the group Γ^m of diffeomorphisms of S^{m-1} modulo those which extend over D^m.

In Section 4 we discuss tubular neighborhoods of neat submanifolds and of submanifolds of the boundary, and in Section 5 the special case of tubular neighborhoods of inverse images of regular values.

1 Exponential Map

Consider the total space E of a smooth vector bundle ξ with base N, a smooth manifold, and projection π. We identify N with the zero section of E and note that every open neighborhood U of N in E contains a neighborhood that is a total space of a bundle: We can shrink E to a subset of U. We want to show that in quite general situations submanifolds possess such bundle neighborhoods. For this reason we begin with a closer examination of the special case of the zero section N of the bundle E.

We will look first at the tangent bundle TN as a subbundle of $T_N E$, the restriction of TE to N. Since $\pi | N = \mathrm{id}$, $D\pi$ is surjective on $T_N E$ and $T_N E = \mathrm{Ker}(D\pi) \oplus TN$. If E_p is the fiber of E at p, then its tangent space at $\mathbf{0}$ is contained in $\mathrm{Ker}(D_p \pi)$, thus, for dimensional reasons, must equal it. This justifies the name for $\mathrm{Ker}(D\pi)$: the *bundle tangent to fibers*. This bundle is actually isomorphic to ξ. To see this recall that the exponential map $T_0 \mathbf{R}^n \to \mathbf{R}^n$ was defined in I,4.5 in an invariant way and was equivariant with respect to linear maps of $\mathbf{R}^n$. Thus, we can define a map $\exp_{\mathrm{Ker}} \colon \mathrm{Ker}(D\pi) \to E$ by requiring it to be on each fiber the exponential map $\mathrm{Ker}_p(D\pi) = T_0 E_p \to E_p$. Collecting all this, we have:

(1.1) Proposition $T_N E = \mathrm{Ker}(D\pi) \oplus TN$; $\mathrm{Ker}(D\pi) \simeq \xi$, *its fibers are tangent spaces to fibers of ξ.*   □

Exercise   Show that if N is an odd dimensional sphere and E its tangent bundle, then $T_N E$ is a trivial bundle.

To progress further we have to generalize the notion of the exponential map $T_0 \mathbf{R}^n \to \mathbf{R}^n$ to a map of the tangent bundle of an arbitrary manifold N to N. In the restricted situation of I,4.5 the exponential mapped a vector v in $T_0 \mathbf{R}^n$ to a vector in the line tangent at $\mathbf{0}$ to v. The generalization will consist in replacing the line in $\mathbf{R}^n$ tangent at $\mathbf{0}$ to v by a curve γ_v in N tangent to v. The appropriate curve is the geodesic, which means that the construction requires a choice of Riemannian metric on N. Once this choice is made, and assuming N to be closed manifold, we have the following fundamental theorem:

(1.2) Theorem   *There is a neighborhood U of N in TN and a smooth map of $U \times I$ to N, $(v, t) \mapsto \gamma_v(t)$, such that*:

   (a) $\gamma_v(0) = \pi(v)$ and the tangent vector of γ_v at 0 is v;
   (b) $\gamma_{sv}(t) = \gamma_v(st)$, $v \in TN$, $s, t \in I$.

The proof of this theorem consists in showing that the curves γ_v are solutions of a certain system of second order differential equations. By an appropriate modification of the differential equations involved this can be deduced from I,6.1. We will not give the details here. They can be found, e.g., in [M1].

With suitable assumptions about the Riemannian metric 1.2 is valid for manifolds with boundary as well.

We can now define the exponential map in a general context.

(1.3) Definition   exp: $U \to N$ is defined by $\exp(v) = \gamma_v(1)$ (see Fig. III,1). (If N is compact, then one can take $U = TN$.)

Consider now the differential of exp restricted to the zero section of TN; this is a map $T_N TN \to TN$. By 1.1

(1.3.1) $T_N TN = \mathrm{Ker}(D\pi) \oplus TN.$

The following lemma describes the behavior of D exp on each summand:

(1.4) Lemma $D \exp | \mathrm{Ker}(D\pi) = \exp_{\mathrm{Ker}}, \ D \exp | TN = \mathrm{id}.$

Proof Fix a vector v in $T_p N$ and let L be the line in $T_p N$ that is the set of all multiples of v. By 1.2(b) $\exp(sv) = \gamma_v(s)$; hence exp maps L onto the geodesic γ_v. Therefore D exp maps the vector in $T_0(T_p N)$ tangent to L to the vector tangent to γ_v at p, that is, the vector v by 1.2(a). Since $T_0(T_p N)$ is the fiber of $\mathrm{Ker}(D\pi)$, this is precisely the description of $\exp_{\mathrm{Ker}} | \mathrm{Ker}(D\pi)$.

Since exp is the identity map on the zero section $(=N)$ of TN, so also is its differential on the tangent bundle to the zero section.   □

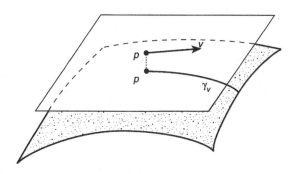

Figure III,1.

2   Normal Bundle and Tubular Neighborhoods

Suppose that a manifold M (with or without boundary) is a submanifold of a manifold N. In the search for a fiber bundle which is a neighborhood of M the first object to find is an appropriate bundle structure. The answer is suggested by 1.1.

(2.1)   *Definition*   The *normal bundle* of M in N, denoted νM, is the quotient bundle $T_M N / TM$. If $f: M \to N$ is an imbedding, then the normal bundle to f, ν_f, is defined by $\nu_f = f^* \nu(fM)$.

(It is possible to modify this definition so as to obtain the normal bundle to an immersion. For if $f: M \to N$ is an immersion, then there is a natural monomorphism $g: TM \to f^*(TN)$ such that the diagram

$$
\begin{array}{ccc}
TM & \xrightarrow{\ Df\ } & TN \\
& \searrow\ g \quad \nearrow & \\
& f^*(TN) &
\end{array}
$$

is commutative. Then $g(TM)$ is a subbundle of $f^*(TN)$ and we define the normal bundle to the immersion f as the quotient $f^*(TN)/g(TM)$.)

Exercise   Show that the normal bundle to $S^n \subset \mathbf{R}^{n+1}$ is trivial.

Exercise   Show that if M is a submanifold of N and N a submanifold of W, then $\nu^W M = \nu^N M \oplus \nu^W N \mid M$. ($\nu^W M =$ the normal bundle of M in W.)

The definition of the normal bundle does not involve a Riemannian structure on N or M. However, we want to represent νM as a subbundle of $T_M N$ and for this we will need the Riemannian structure. Therefore we will assume again that N has a Riemannian metric and represent νM as the complementary bundle to TM in $T_M N$. νM is an $(n-m)$-dimensional vector bundle, thus an n-dimensional manifold. We will show that it is the desired fibration of a neighborhood of M. For this purpose we assume that $\partial N = \emptyset$, so that exp is defined on a neighborhood of the zero section ν_0 of νM and maps it into M.

(2.2)   **Theorem**   *Suppose that M is a closed subset of N, $\partial N = \emptyset$. Then there is a neighborhood of ν_0 on which exp is an imbedding.*

Proof First, we will show that $\exp | \nu$ is of maximal rank on ν_0. Fix $p \in M$ and consider T_pN. We have

(*) $$T_pN = \nu_p \oplus T_pM,$$

where ν_p is the fiber of ν at p. Also, by 1.3.1,

(**) $$T_pTN = \operatorname{Ker}_p D\pi \oplus T_pN,$$

where $\operatorname{Ker}_p D\pi$ is the tangent space to T_pN (see Fig. III,2).

Now, consider $T_p\nu$. Since ν is a subbundle of T_MN, $T_p\nu \subset T_pTN$. By 1.1

(***) $$T_p\nu = \operatorname{Ker}_p D\pi_\nu \oplus T_pM, \ \pi_\nu = \pi | \nu,$$

where $\operatorname{Ker}_p D\pi_\nu$ is the tangent space to ν_p. Thus $\operatorname{Ker}_p D\pi_\nu$ is contained in the first factor of the direct sum in (**) and T_pM in the second factor. It follows from 1.4 that $D \exp | T_p\nu$ maps $\operatorname{Ker}_p D\pi_\nu$, the first factor in (***), isomorphically onto ν_p, the first factor in (*), and is an identity on T_pM, the second factors in both. Thus, it is an isomorphism.

Since $\exp | \nu$ is of maximal rank on ν_0, it is of maximal rank in a neighborhood U of ν_0, i.e., it is an immersion on U. It is a homeomorphism—the identity map—on ν_0. Since M is a closed subset of N, I,7.2 applies and implies that $\exp$ is a homeomorphism on a neighborhood V of M. Thus it is an imbedding on $U \cap V$. $\square$

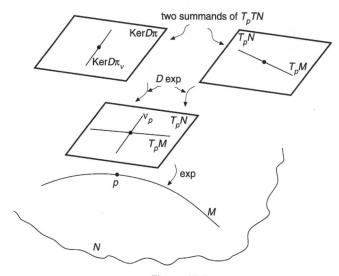

Figure III,2.

(2.3) Corollary *Suppose that $f\colon M \to N$ imbeds M as a closed subset of N. Then f extends to an imbedding $\bar{f}$ of ν_f in N. If $\partial M = \emptyset$, then $\bar{f}(\nu_f)$ is an open neighborhood of $f(M)$ in N.*

Proof By 2.2 there is a neighborhood U of M in ν_f and an extension of f to an imbedding f' of U. Now, if $g\colon \nu_f \to \nu_f$ is a diffeomorphism which shrinks ν_f so that $g(\nu_f) \subset U$ (*cf.* I,3.5), then $\bar{f} = f'g$ is the desired imbedding. If M is a closed manifold, so also is ν_f and, by the Invariance of Domain, $\bar{f}(\nu_f)$ is an open subset of N containing $f(M)$. $\square$

We now define tubular neighborhoods. As before, M is assumed to be a submanifold of N, dim $M = m$, dim $N = n$.

(2.4) *Definition* A subset of N which has the structure of a $(n - m)$-dimensional vector bundle over M with M as the zero section is called a *tubular neighborhood* of M. A subset of N which has the structure of an $(n - m)$-disc bundle over M with M as the zero section is called a *closed tubular neighborhood* of M.

It follows from 2.3 that if M is a closed subset of N and $\partial N = \emptyset$, then M possesses a tubular neighborhood with the vector bundle structure that of the normal bundle. We will show in 3.1 that this is the only vector bundle structure possible in a tubular neighborhood.

We do not require that a tubular neighborhood be an open neighborhood. However, if M has no boundary, then its tubular neighborhood is a submanifold of N of the same dimension as N and without boundary. Hence it is an open neighborhood of M. If the boundary of M is not empty, then its tubular neighborhood is not an open subset of N.

If the boundary of N is not empty and touches the interior of M, then M does not have a tubular neighborhood in N. A satisfactory theory in this case is obtained for neat submanifolds of N. We will do this in Section 4.

Let F be a tubular neighborhood with a Riemannian structure, and let ε be a smooth positive function on M. Then the set of all vectors v in F such that the length of v is $\leq \varepsilon(p)$ if $v \in F_p$ is a closed tubular neighborhood of M. Its interior is an open disc bundle and it is also the result of an ε-shrinking of F (*cf.* I,3.5).

The converse also holds: A closed tubular neighborhood of a compact submanifold M, which is a closed neighborhood of M in N, can always be realized as a closed disc subbundle of a tubular neighborhood of M. For if E is such a neighborhood, then we first reparametrize its interior to make it a vector bundle and then consider the unit disc subbundle E'. E' can be expanded by an isotopy to cover E and since it is compact this isotopy can be extended to an isotopy of N. The resulting isotopy will expand the interior of E to a tubular neighborhood of M containing E as a closed disc subbundle.

Exercise Show that if U is an open neighborhood of M in N and M has a tubular neighborhood, then it has one contained in U.

The very fact of the existence of tubular neighborhoods allows an easy proof of the following theorem stating, roughly, that in the realm of the homotopy theory of smooth manifolds, continuous maps can always be replaced by smooth ones.

(2.5) Theorem *Let M and N be smooth manifolds, and $d(p, q)$ a metric on N. Assume that ∂N is compact. Then:*

(a) *There is a continuous positive function δ on N such that if $f, g \colon M \to N$ are two continuous maps and $d(f(p), g(p)) < \delta(f(p))$, then f is homotopic to g;*

(b) *If $f \colon M \to N$ is a continuous map smooth on a closed subset $K \subset M$ and ε is a continuous positive function on N, then there is a smooth map $g \colon M \to N$ such that $f = g$ on K and $d(f(p), g(p)) < \varepsilon(f(p))$.*

Proof Assume first that $\partial N = \varnothing$ and that N is a closed submanifold of $\mathbf{R}^m$. Since the metric on N induced from $\mathbf{R}^m$ and the metric $d(p, q)$ induce the same topology, the theorem will follow if we prove it with $d(p, q)$ interpreted as the former.

To prove (a) let F be a tubular neighborhood of N in $\mathbf{R}^m$ and $f, g \colon M \to N$ two continuous maps. Since F is an open neighborhood of N one shows easily, using partitions of unity, that there is a continuous function δ on N such that if $p, q \in N$ and $|p - q| < \delta(p)$, then the straight line segment pq is in F. Thus if $|f(p) - g(p)| < \delta(f(p))$, then f and g are homotopic as maps into F. Composing this homotopy with the bundle projection $\pi \colon F \to N$ shows they are homotopic as maps into N.

To prove (b) note first that there is a tubular neighborhood F of N in $\mathbf{R}^m$ such that $d(p, \pi(p)) < \varepsilon(\pi(p))/2$ in the metric on $\mathbf{R}^m$. Now, I,2.5

applied to the map f viewed as a map to $\mathbf{R}^m$ yields a smooth map $g': M \to F$ such that $d(f(p), g'(p)) < \varepsilon(f(p))/2$. Then $g = \pi g'$ is as desired.

If N has a non-empty boundary, then shrink N to $N' \subset \text{Int } N$ and note that the tubular neighborhood of Int N is an open neighborhood of N'. Therefore one can proceed as before.

Finally, N can always be assumed to be a closed submanifold of $\mathbf{R}^m$ for some m. We have shown this in II,3.1 for a compact N; for the general case see [Wi2]. □

(2.6) Corollary *If $f, g: M \to N$ are smooth and homotopic (as continuous maps), then they are smoothly homotopic.* □

Exercise Show that if two smooth vector bundles over M are (continuously) isomorphic, then they are smoothly isomorphic.

We close this section with a remark concerning the behavior of a normal bundle during an isotopy. Suppose then that $F: M \times \mathbf{R} \to N$ is an isotopy of an imbedding f.

(2.7) Proposition *There is a bundle ν over $M \times \mathbf{R}$ such that its restriction ν_t to $M \times \{t\}$ is the normal bundle to the imbedding $F_t = F \mid M \times \{t\}$.*

Proof Let $\pi_M: M \times \mathbf{R} \to \mathbf{R}$, $\pi_N: N \times \mathbf{R} \to \mathbf{R}$ be the projections, and let $T_h(M \times \mathbf{R}) = \text{Ker } D\pi_M$, $T_h(N \times \mathbf{R}) = \text{Ker } D\pi_N$. Observe that $T_h(M \times \mathbf{R}) \mid M \times \{t\}$ is the tangent bundle to $M \times \{t\}$. If $G: M \times \mathbf{R} \to N \times \mathbf{R}$ is the level preserving imbedding associated to F, then $DG(T_h(M \times \mathbf{R}))$ is a subbundle of $T_h(N \times \mathbf{R})$ and $\nu = G^*(T_h(N \times \mathbf{R})/DG(T_h(M \times \mathbf{R})))$ is as desired. (The bundle $T_h(N \times \mathbf{R})/DG(T_h(M \times \mathbf{R}))$ is the horizontal normal bundle of the submanifold $G(M \times \mathbf{R})$.) □

Observe that $\nu_0 = \nu_f$ and that, as for every bundle over a product with $\mathbf{R}$, $\nu = \pi^* \nu_0$ where $\pi: M \times \mathbf{R} \to M \times \{0\}$ is the projection. Therefore $\nu = \pi^* \nu_f$, which implies the following corollary:

(2.8) Corollary *If $f, g: M \to N$ are two isotopic imbeddings, then $\nu_f \cong \nu_g$.* □

Of course, similar results hold for the normal bundle to an immersion with isotopy replaced by homotopy through immersions.

3 Uniqueness of Tubular Neighborhoods

The uniqueness of vector bundle structure of tubular neighborhoods is a consequence of the following theorem.

(3.1) **Theorem** *Let M be a closed submanifold of N and $F^0 \subset N$ the total space of a k-dimensional vector bundle with M as the base. Let F^1 be a tubular neighborhood of M. Then there is an isotopy G_t of the inclusion $F^0 \subset N$ such that $G_t(p) = p$ for $p \in M$ and G_1 is a linear map $F^0 \to F^1$ of rank k on each fiber.*

Proof We will say that $U \subset M$ is a trivializing chart for a bundle F if U is a chart in M and the restriction F_U is trivial.

We will show first that F^0 can be ε-shrunk to a bundle E^0 with the following property:

(*) For every $p \in M$ there is a neighborhood U in M and a trivializing chart V for F^1 such that $E_U^0 \subset F_V^1$.

To see that this is so, let V be a trivializing chart for F^1 and let $p \in V$. Since F_V^1 is an open neighborhood of p, there are a number ε_p and a neighborhood U of p in M such that all ε_p-discs in the fibers of F^0 over U are contained in F_V^1. We can assume that such sets U form a locally finite covering $\{U_\alpha\}$ of M with corresponding functions ε_α. Using the associated partition of unity we can construct a smooth positive function ε on M with the property that for every $p \in M$, $\varepsilon(p) < \varepsilon_\alpha$ for some α. Then the ε-shrinking of F^0 results in a bundle E^0 with property (*).

Clearly, the shrinking map is isotopic to the inclusion $F^0 \to N$. Thus, to conclude the proof, it is enough to construct an isotopy of the inclusion $\iota: E^0 \hookrightarrow F^1$ to a linear non-degenerate map $E^0 \to F^1$. We do it as follows:

For $t \neq 0$, we define the desired isotopy G_t by

$$G_t(v) = \frac{1}{t}\iota(tv).$$

To define G_t for $t = 0$, note first that if U, V are as in (*) then the inclusion $\iota: E_U^0 \subset F_V^1$ is given in local coordinates by a map

$$v = (x, y) \mapsto (f(x, y), g(x, y)) \in \mathbf{R}^m \times \mathbf{R}^{n-m}, \qquad x \in \mathbf{R}^m, y \in \mathbf{R}^k,$$

where $f(x, 0) = x$, $g(x, 0) = 0$.

Writing $g = (g_1, \ldots, g_{n-m})$, we have, by A,2.1,

$$g_i(x, y) = \sum_j a_j^i(x, y)y_j, \qquad \text{where } a_j^i(x, \mathbf{0}) = \frac{\partial g_i}{\partial y_j}(x, \mathbf{0}),$$

$i = 1, \ldots, n - m, \; j = 1, \ldots, k.$ Therefore,

$$G_t(x, y) = (f(x, ty), \frac{1}{t} g(x, ty))$$

$$= (f(x, ty), \sum_j a_j^1(x, ty)y_j, \ldots, \sum_j a_j^{n-m}(x, ty)y_j),$$

which is smooth and well-defined for all t.

Now, G_0 maps the fiber of E^0 at x into the fiber of F^1 at x by a linear map given by the matrix $J = (a_j^i(x, \mathbf{0}))$. To calculate the rank of J we observe that the Jacobian $J(\iota)$ is of rank $m + k$ and that along M, i.e., for $y = \mathbf{0}$,

$$J(\iota) = \begin{pmatrix} I_m & * \\ 0 & J \end{pmatrix}$$

Thus J must be of rank k, which shows that G_0 is as desired. $\square$

Theorem 3.1 applies to the case where M is a closed manifold and F^0, F^1 are two tubular neighborhoods of M in N. We obtain:

(3.2) Corollary *Every two tubular neighborhoods of a closed submanifold are isomorphic.*

There is also a version of 3.1 that applies to collars. To see this, let $M = \partial N$ and let F^0, F^1 be two collars of M. We view F^0, F^1 as diffeomorphisms of $M \times \mathbf{R}_+$ into N.

(3.3) Theorem F^0 *is isotopic to* F^1 *by an isotopy that is stationary on* M.

Proof Since multiplication by non-negative numbers is allowed in $\mathbf{R}_+$, the proof of 3.1 goes without change and yields an isotopy $G_t(p, s), (p, s) \in M \times \mathbf{R}_+, t \in I$, such that $G_1 = F^1$, $G_0(p, s) = F^0(p, a(p)s)$ for some smooth positive function $a(p)$ on M. This last map is clearly isotopic to F^0. $\square$

Simple examples show that one cannot in general require the isotopy in 3.1 to be ambient. Even in the case when both F^0 and F^1 are tubular neighborhoods of the same manifold M, an additional restriction is necessary.

(3.4) *Definition* A tubular neighborhood is called *proper* if it is obtained by ε-shrinking of another tubular neighborhood, which sends fibers onto fibers.

It is easy to see that the interior of a closed tubular neighborhood of M is a proper tubular neighborhood and that every proper tubular neighborhood is the interior of a closed tubular neighborhood.

A tubular neighborhood is a neighborhood with a definite vector bundle structure on it. Accordingly, it may fail to be proper in two ways: as a set, and as a bundle. For instance, $\mathbf{R}^m$ is a tubular neighborhood of the origin $\mathbf{0}$ that no bundle structure could make proper. On the other hand, the strip $|x| < \pi/2$ in $\mathbf{R}^2$ is a proper tubular neighborhood of the line $x = 0$ when fibered by segments $y = $ const, but not a proper one when fibered by curves $y = \tan x + $ const.

For proper tubular neighborhoods of compact submanifolds we have the Tubular Neighborhood Theorem:

(3.5) **Theorem** *If M is compact and closed and if F^0, F^1 are either proper or closed tubular neighborhoods of M in N, then there is an isotopy H_t of the identity map of N that keeps M fixed and such that $H_1|F^0$ is an isometry $F_0 \to F_1$.*

Proof Assume first that F^0, F^1 are proper tubular neighborhoods obtained by shrinking tubular neighborhoods E^0, E^1 to the unit disc bundles. By 3.1 there is an isotopy H_t of the inclusion $E^0 \subset N$ to an isomorphism $E^0 \to E^1$. By II,4.6 there is an isotopy G_t of the isomorphism H_1 to an isometry G_1. Let K be H followed by G. By II,5.2, K restricted to the unit disc bundle of E^0 extends to an isotopy of the identity map of N.

If F^0 and F^1 are closed tubular neighborhoods, then they are unit disc bundles of proper tubular neighborhoods E^0, E^1, and the theorem follows from the case already considered. $\square$

A similar argument, left to the reader, shows that the isotopy in 3.3 can be assumed to be ambient—provided that ∂N is compact and the collars proper in the obvious sense. This result is known as *uniqueness of the collars.*

The following corollary due to R. Palais [Pa1], is known as the Disc Theorem.

(3.6)   Corollary   *Let f, g be two imbeddings of the closed disc D^k in the interior of a connected manifold N^n. If $k = n$, then assume that f and g are either both orientation preserving or both orientation reversing (i.e., are equioriented). Then f is ambient isotopic to g.*

If $f = g$ on a disc $D^m \subset D^k$, then this isotopy may be assumed to be stationary on D^m.

The same is true if f and g are imbeddings of the open disc, provided that both extend to imbeddings of $\mathbf{R}^k$ in N.

Proof   As in the proof of 3.5 we prove this for imbeddings of the open disc; the other case follows from this.

By II,5.3 there is an isotopy H_t of the identity map of N such that $H_1(f(0)) = g(0)$. But then, if $k = n$, $H_1(f(D))$ and $g(D)$ are both proper tubular neighborhoods of $g(0)$ and the corollary follows from 3.5 and II,4.5. If $k < n$ and f, g extend to imbeddings of $\mathbf{R}^k$, then they extend to imbeddings of $\mathbf{R}^n$ (take tubular neighborhoods) and the corollary follows from the case $k = n$.   □

For future reference we list here another version of the Disc Theorem. The inductive proof is left as an exercise.

(3.7)   Corollary.   *Let mD^k be the disjoint union of m copies of the closed disc D^k and f, g two imbeddings of mD^k in a connected manifold N^n. If $k = n$ then we assume that f and g are equioriented. Then f is isotopic to g.*   □

Exercise   Suppose that $f, g: M^m \to S^n$, $m < n$, are two imbeddings of a compact manifold M^m and that $Hf = g$ for some orientation preserving diffeomorphism $H: S^n \to S^n$. Show that f and g are isotopic. (*Hint*: $f(M)$ is contained in an imbedded disc.)

4   Submanifolds of the Boundary

In 2.3, we showed that submanifolds situated in the interior of a given manifold N possess tubular neighborhoods. Now, we will show that the same holds for neat submanifolds of a manifold with boundary.

Let N be a manifold with boundary and let ν be a subbundle of $T_{\partial N}N$ spanned by a vector field X pointing inside N. Then $T_{\partial N}N = T\,\partial N \oplus \nu$. Now, if $M \subset N$ is a neat submanifold, then, by II,2.8, we can assume that, along M, X points inside M. Then $T_{\partial M}M = T\,\partial M \oplus \nu$ and we have a natural identification

$$T_{\partial M}N/T_{\partial M}M = T\,\partial N/T\,\partial M$$

of bundles restricted to ∂M. In other words, we can identify the normal bundle of M restricted to ∂M with the normal bundle of ∂M in ∂N. This explains and justifies the following definition.

(4.1) Definition Let F be a tubular neighborhood of a neat submanifold M of the manifold N. We say that F is *neat* if $F \cap \partial N$ is a tubular neighborhood of ∂M in ∂N. This implies that F is an open subset of N.

(4.2) Theorem *If M is a neat submanifold of N, then it has a neat tubular neighborhood.*

Proof According to I,7.5, ∂N has a product neighborhood $\partial N \times \mathbf{R}_+$ in N. Therefore there is a Riemannian metric on N that is a product metric in this neighborhood. This implies that, if $\nu M|\partial N$ is identified with the normal bundle of ∂M in ∂N, then the geodesics corresponding to normal vectors of ∂M and issued at points of ∂N will stay in ∂N. Therefore the construction employed in the proof of 2.2, unchanged, will yield a neat tubular neighborhood of M in N.   □

The Tubular Neighborhood Theorem remains valid for neat proper neighborhoods of compact submanifolds. The proof is left as an exercise.

We will extend now the definition of tubular neighborhoods so as to apply to submanifolds of the boundary. They do not possess tubular neighborhoods in the sense of 2.4. However, they have "half-tube" neighborhoods. For instance, if $M = \partial N$, then it is reasonable to consider a collar of M in N as its tubular neighborhood. The normal bundle of M in N is trivial, i.e., it can be identified with $M \times \mathbf{R}$, and a collar is an imbedding of $M \times \mathbf{R}_+$. In this context, the Uniqueness of Collars Theorem from the last section plays the role of 3.5.

Now, if M is a submanifold of $B = \partial N$, then its normal bundle in N is the Whitney sum of the normal bundle $\nu^B M$ of M in B and of the normal bundle of B in N, which is trivial. Its total space is then $\nu^B M \times \mathbf{R}$. We define now the tubular neighborhood of M in N to be an imbedding of $\nu^B M \times \mathbf{R}_+$ in N extending an imbedding of $\nu^B M$ in B as a tubular neighborhood of M in B. Clearly, if M has a tubular neighborhood in B then such an imbedding can always be constructed using the collar of B.

Definition 3.4 of proper tubular neighborhood remains unchanged. The Tubular Neighborhood Theorem is still valid; the isotopies are compositions of the isotopies of $\nu^B M$ and of the isotopies of the collar.

As an example, and for use later, we construct a tubular neighborhood of a section of a sphere bundle in the bundle itself and in the associated unit disc bundle.

Let ξ be a Riemannian vector bundle over M with total space E. The set S of vectors of length one in E forms a fiber bundle over M, the unit sphere bundle. Let $s: M \to S$ be a section. Then, s is a nonvanishing section of ξ and spans a trivial line bundle η. Let $\eta^\perp$ be the complementary bundle with total space F. Then $\eta^\perp \oplus \eta = \xi$. We will show that $\eta^\perp$ is isomorphic to the normal bundle to $s(M)$ in S. To see this it is enough by 3.2 to imbed the total space F of $\eta^\perp$ as a tubular neighborhood of s. Such an imbedding can be constructed by wrapping each fiber F_x on the corresponding fiber of S by the projection from $-s(x)$. That is, we define $p: F \to S$ by

$$(4.3) \qquad p(v) = \frac{2}{1 + v^2} v + \frac{1 - v^2}{1 + v^2} s(x)$$

for $v \in F_x$ (*cf.* I,1.2).

Then p is a diffeomorphism onto $S - (-s(M))$ and $p \,|\, M = s$, as required.

If we compose p with an ε-shrinking of F we obtain a proper tubular neighborhood of $s(M)$. Note that for $\varepsilon = 1$ the corresponding tubular neighborhood consists of "one-half" of S, that is, its intersection with every sphere S_x is that hemisphere of S_x which contains $s(x)$.

The set D of vectors of length ≤ 1 in E is the unit disc bundle. It is a manifold with boundary, $\partial D = S$. To find a tubular neighborhood of s in D we have to extend 4.3 to an imbedding of $F \times \mathbf{R}_+$ in D. This is done conveniently by setting

$$(4.4) \qquad p(v, t) = \frac{2}{v^2 + (1 + t)^2} v + \frac{1 - v^2 - t^2}{v^2 + (1 + t)^2} s(x)$$

for $(v, t) \in F_x \times \mathbf{R}_+$.

Again, this is not a proper neighborhood: It consists of the entire bundle D with the antipodal section $-s(M)$ deleted.

The geometric meaning of 4.4 is explained in VI,3.3.

5 Inverse Image of a Regular Value

Let M be a closed manifold, $f: M \to N$ a smooth map, and p a regular value of f. Then $V = f^{-1}(p)$ is a submanifold of M and $TV \subset \operatorname{Ker} Df$. Since Df is surjective, $\dim \operatorname{Ker} Df = \dim M - \dim N = \dim V$; hence $TV = \operatorname{Ker} Df$. Thus Df induces a map $\nu V = T_V M / TV \to T_p N$, which is an isomorphism on each fiber. This means that $\nu V = (Df)^* T_p N$ is a trivial bundle, and tubular neighborhoods of V are product neighborhoods. In general, the map f restricted to such a tubular neighborhood cannot be identified with a projection. However, if V is compact, then it admits tubular neighborhoods with this property.

(5.1) Proposition *If V is compact, then there is a neighborhood U of p in N and an imbedding $j: V \times U \to M$ such that the diagram*

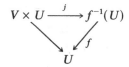

commutes.

Proof Let W be a chart about p and let $V \times \mathbf{R}^n \subset M$ be a tubular neighborhood of V; we identify W with $\mathbf{R}^n$ and assume that $f(V \times \mathbf{R}^n) \subset W$. To prove the proposition it is enough to show that:

(∗) There is a neighborhood U of p such that for every $q \in V$ the map f restricted to $U_q = f^{-1}(U) \cap \{q\} \times \mathbf{R}^n$ is a diffeomorphism of U_q onto U.

For if this is the case and $g_q: U \to U_q$ is the inverse of $f \mid U_q$ then the imbedding $j: V \times U \to M$ given by $j(q, v) = g_q(v)$ satisfies $fj(q, v) = fg_q(v) = v$.

To prove (∗), consider the map $h: V \times \mathbf{R}^n \to V \times \mathbf{R}^n$, $h(q, v) = (q, f(q, v))$. For a given q this map is of maximal rank at $(q, \mathbf{0})$; hence there exists a neighborhood V_q of q and W_q of p such that $h \mid V_q \times W_q$ is a diffeomorphism onto $h(V_q \times W_q)$ (see Fig. III,3). This implies that f restricted to $\{r\} \times W_q$ is a diffeomorphism onto $f(W_q)$ for all r in V_q. Since V is compact there is a finite family $V_1 \times W_1, \ldots, V_k \times W_k$ such that $V = \bigcup_i V_i$. Let $U' \subset \bigcap_i W_i$ be an open neighborhood of p and let $U \subset U'$ be another neighborhood satisfying $f^{-1}(U) \subset V \times U'$; that such U can be found follows from the

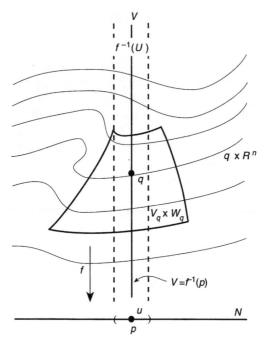

Figure III,3.

compactness of V and II,2.3. Since $f|\{q\} \times U'$ is a diffeomorphism and $f^{-1}(U) \cap \{q\} \times \mathbf{R}^n \subset \{q\} \times U'$, U satisfies (∗).   □

Exercise   Assume that M is closed, connected, and compact, and that N is connected. Show that if there is a submersion $f\colon M \to N$, then N is closed, $f(M) = N$, and M is a fiber bundle over N with the projection f. (This is due to C. Ehresmann [Eh].)

6   The Group Γ^m

Let M be an oriented manifold, possibly with boundary. The set Diff M of orientation preserving diffeomorphisms of M onto itself is a group under the operation of composition. It is a very large group, non-abelian except in trivial cases. We shall use 3.6 to study a certain quotient of Diff S^{m-1} of special importance.

Let $\text{Diff}_0\, S^{m-1}$ consist of those diffeomorphisms of S^{m-1} which are isotopic to identity. Since isotopies can be composed, this is a subgroup.

(6.1)   Lemma $\text{Diff}_0\, S^{m-1}$ *contains the commutator subgroup of* $\text{Diff}\, S^{m-1}$.

Proof   Let f and g be two orientation preserving diffeomorphisms of S^{m-1}. It follows from 3.6 that there are diffeomorphisms f_+ and g_- isotopic to, respectively, f and g, and such that f_+ is an identity on the northern hemisphere of S^{m-1} and g_- on the southern hemisphere. Now, the commutator of f and g is isotopic to the commutator of f_+ and g_-. Since f_+ and g_- commute, the latter is the identity map of S^{m-1}.   □

Consider now those diffeomorphisms of S^{m-1} that can be extended over D^m, i.e., the image of $\text{Diff}\, D^m$ under the restriction homomorphism $\partial: \text{Diff}\, D^m \to \text{Diff}\, S^{m-1}$. Since $\text{Diff}_0\, S^{m-1} \subset \partial\, \text{Diff}\, D^m$, 6.1 implies that both are normal subgroups of $\text{Diff}\, S^{m-1}$ with abelian quotient. Letting $\Gamma^m = \text{Diff}\, S^{m-1}/\partial\, \text{Diff}\, D^m$, we have:

(6.2)   Proposition   *The group* Γ^m *of the diffeomorphisms of* S^{m-1} *modulo those which extend over* D^m *is an abelian group.*   □

We will show in VIII,5.6 that, for $m > 4$, Γ^m is in a 1-1 correspondence with the set of distinct differentiable structures on the sphere S^m. For $m \leq 4$ the groups Γ^m vanish: That $\Gamma^2 = 0$ was the subject of the exercise in II,4.6; that $\Gamma^3 = 0$ is due to Smale [Sm1] and Munkres [Mu1]. Finally, Cerf proved that $\Gamma^4 = 0$. The proof occupies 132 pages [C2].

7   Remarks

The idea of a normal vector is as old as differential geometry itself, but the notion of a tubular neighborhood of a submanifold as a neighborhood fibered by normal planes emerged much later in the work of H. Whitney, [Wi1] and [Wi2, Section 28]. [Wi1] is perhaps the earliest paper considering fiber bundles; the normal sphere bundle is one of the examples considered there.

A combinatorial analogue of tubular neighborhoods was introduced in 1938 by J. H. C. Whitehead. In [Wh2] he defined a regular neighborhood of a subcomplex K of a combinatorial manifold as a submanifold which contracts geometrically into K, and proved that different regular neighborhoods of the same subcomplex are combinatorially equivalent. In turn, a differential analogue of this is, of course, the Tubular Neighborhood Theorem. It was proved in 1961 by J. Milnor in his course at Princeton University and by C. T. C. Wall in a seminar at Cambridge.

This interplay of ideas between differential and combinatorial topology can be seen also in other places. For instance, the old notion of general position was used in topology for similar purposes as the notion of transversality introduced much more recently into differential topology by Thom.

IV

Transversality

The notion of transversality is a smooth equivalent of the notion of general position. For instance, two submanifolds M^m and V^r of N^n, $n \leq m + r$, are transversal if their intersection looks locally like the intersection in $\mathbf{R}^n$ of the subspace of the first m coordinates with the subspace of the last r coordinates. This geometric idea is properly expressed as transversality of maps and defined in terms of their differentials. This is done in Section 1.

The ability of deform maps to a transversal position is one of the most powerful techniques of differential topology. A general theorem in this direction is given here in 2.1; it will be in constant use in subsequent chapters.

In Sections 3 and 4 we apply transversality to establish foundations of Morse theory of critical points of differentiable functions. In Section 5 we use it to define intersection numbers.

1 Transversal Maps and Manifolds

(1.1) *Definition* Let $f: M \to N$, $g: V \to N$ be two smooth maps. We say that f is *transversal* to g, $f \pitchfork g$, if whenever $f(p) = g(q)$, then $Df(T_p M) + Dg(T_q V) = T_{f(p)} N$.

59

Note that this condition is equivalent to the requirement that the composition

(1.2) $$T_pM \xrightarrow{Df} T_{f(p)}N \longrightarrow T_{f(p)}N/Dg(T_qV)$$

be surjective.

Obviously, if dim M + dim V < dim N, then $f \pitchfork g$ is possible only if $f(M)$ and $g(V)$ are disjoint.

The notation $f \pitchfork g$ will be replaced by $f \pitchfork V$ whenever V is a submanifold and g an identity map. The meaning of $M \pitchfork V$ is also clear.

In certain situations the second map in 1.2 is a differential of a map; hence the composition is also a differential. This is the case when V is a fibre of a smooth fibre bundle N with projection π. Then, if f maps a manifold M into N, the differential of πf is precisely the composition in 1.2. This differential is surjective if and only if the point $\pi(V)$ is a regular value of πf. Thus we have:

(1.3) Proposition *Let $f: M \to N$, where N is a smooth fiber bundle with projection π, and let F_q be a fiber over a point q. Then $f \pitchfork F_q$ if and only if q is regular value of πf.* $\square$

Viewing the product $W \times V$ as a bundle over W, we obtain from this and the Brown–Sard Theorem (A,3.1) the following:

(1.3.1) Corollary *If $f: M \to W \times V$, then there is a dense set of points $q \in V$ such that $f \pitchfork W \times \{q\}$.* $\square$

As another corollary we have a characterization of cross sections:

(1.3.2) Corollary *Let N be a smooth fiber bundle over M. A submanifold $V \subset N$ is a cross section of the bundle if and only if V intersects every fiber F_q transversely in a single point $s(q)$.*

Proof The necessity is clear. To prove that the condition is sufficient we have to show that the map $s: M \to N$ is smooth. To do this, we first note that s is the inverse of $\pi|V$ and that, by 1.3, $D(\pi|V): T_{s(q)}V \to T_qM$ is surjective. Since dim V = dim M, $D(\pi|V)$ is an isomorphism. Now, it follows from the Implicit Function Theorem (A,1.1) that the inverse of $\pi|V$ is smooth. $\square$

For example, if V is the image of the imbedding $\mathbf{R} \to \mathbf{R}^2$ given by $t \mapsto (t^3, t)$, then V is a smooth submanifold of $\mathbf{R}^2$ and a continuous section of $\mathbf{R}^2$ considered as a trivial line bundle over the x axis. But it is not a smooth section: It is not transversal to the y axis.

The notion of transversality generalizes that of a regular value: If $f: M \to N$ and $q \in N$, then q is a regular value of f if and only if $f \pitchfork \{q\}$ and $(f|\partial M) \pitchfork \{q\}$. Replacing q by a closed submanifold V, we obtain the following generalization of II,1.7:

(1.4) Proposition *If $f \pitchfork V$ and $(f|\partial M) \pitchfork V$, then $W = f^{-1}(V)$ is a neat submanifold of M. Moreover, $\nu W = f^* \nu V$.*

Proof Let $p \in W$ and $q = f(p)$. By II,2.3(b) there is in N a neighborhood U of q and a map $h: U \to \mathbf{R}^r$ such that $U \cap V = h^{-1}(\mathbf{0})$. Moreover, we can identify Dh at q with $T_q N \to T_q N / T_q V$. Now, $f^{-1}(U)$ is an open neighborhood of $p, f^{-1}(U) \cap W = f^{-1} h^{-1}(\mathbf{0})$, and both Dhf and $D(hf|\partial M)$ are surjective by the assumption. By II,2.3(b) again, W is a submanifold of M.

Note that $\operatorname{codim}_M(W) = \operatorname{codim}_N(V)$.

Let now d be the dimension of the kernel of the composite map

$$T_W M \xrightarrow{Df} T_V N \xrightarrow{\pi} \nu V = T_V N / TV.$$

Since $\pi \circ Df$ is surjective, $m - d \geq \operatorname{codim} V$, i.e., $d \leq m - \operatorname{codim} V = \dim W$. On the other hand, $TW \subset \operatorname{Ker}(\pi \circ Df)$; thus $d \geq \dim W$. It follows that $d = \dim W$; hence $\operatorname{Ker}(\pi \circ Df) = TW$. Therefore $f: W \to V$ induces a bundle map $T_W M / TW = \nu W \to \nu V = T_V N / TV$. $\square$

A very nice application of 1.4 is a simple proof, due to M. Hirsch, of Brouwer's Fixed Point Theorem.

(1.5) Theorem *There is no (continuous) retraction $D^n \to \partial D^n$.*

Proof Observe first that it is enough to prove that there is no smooth retraction. For if $r: D^n \to \partial D^n$ is a continuous retraction, then there is a smooth $1/2$-approximation r' to r that is also the identity map on ∂D^n (*cf.* III,2.5). This is not yet a retraction, but since the origin is not in $r'(D^n)$ we can compose r' with the projection from the origin to obtain a smooth retraction.

Suppose now that $r: D^n \to \partial D^n$ is a smooth retraction, let $p \in \partial D^n$ be a regular value of r, and let L be the connected component of $r^{-1}(p)$ containing

p. Since $r^{-1}(p)$ is a neat submanifold, L is an arc with endpoints p and $q, p \neq q$ and $q \in \partial D^n$. This implies $p = r(q) = q$, a contradiction. $\square$

The notion of transversality already appeared, in disguise, in the definition of neat submanifolds: II,2.8.1 means nothing else but that $M \pitchfork \partial N$. Moreover, as we have seen, this condition characterizes neat submanifolds.

The following theorem, which for simplicity is stated for closed manifolds only, provides the expected geometric justification of the definition of transversality.

(1.6) Theorem *Let M^m and V^r be closed transversal submanifolds of N^n and let $p \in M \cap V$. If $n \leq m + r$, then there is in N a chart U about p in which $U \cap M$ is represented by the space of the first m coordinates and $U \cap V$ is represented by the space of the last r coordinates.*

Proof We will prove this in the special case $\dim N = m + r$. By II,2.3(a) there is a chart U in N about p such that $U \cap M$ corresponds to the space of the first m coordinates. We will simply identify this chart with $\mathbf{R}^m \times \mathbf{R}^r$. The part of V lying in it can then be represented by an image of $\mathbf{R}^r$ under an imbedding $f: \mathbf{R}^r \to \mathbf{R}^m \times \mathbf{R}^r$, where $f(y) = (\alpha(y), \beta(y))$ and $f(0) = 0 = p$. The transversality assumption means that the Jacobian of β is of rank r at 0. Now, consider the map $g: \mathbf{R}^m \times \mathbf{R}^r \to \mathbf{R}^m \times \mathbf{R}^r$ given by

$$g(x, y) = (x + \alpha(y), \beta(y)), \qquad x \in \mathbf{R}^m, y \in \mathbf{R}^r.$$

Note that g at 0 is of rank $m + r$; hence it is a chart if restricted to a suitably small neighborhood U of 0 in $\mathbf{R}^m \times \mathbf{R}^r$. Since $g(0, y) = f(y)$, $g(x, 0) = (x, 0)$, it is precisely the chart we were looking for.

Another proof of this can be based on III,3.1. This method is particularly suitable to the general case. The details are left to the reader. $\square$

(1.7) Corollary *Let M^m, V_1^r, V_2^r be submanifolds of N^n, $n = m + r$. Suppose that V_1, V_2 intersect M in the same point p and that this intersection is transversal. Then there is an isotopy of N that keeps M fixed and brings V_1 to coincide with V_2 in a neighborhood of p.*

Proof By 1.6 there is a chart $U = \mathbf{R}^m \times \mathbf{R}^r$ in N about p that intersects M in $\mathbf{R}^m \times 0$ and V_1 in $0 \times \mathbf{R}^r$ (see Fig. IV,1). A sufficiently small chart $U_2 = \mathbf{R}^r$ about p in V_2 is represented in U as an imbedded $\mathbf{R}^r$ transversal to $\mathbf{R}^m \times 0$ and intersecting it in the origin.

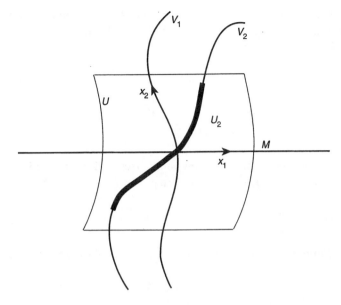

Figure IV,1

Now, we use III,3.1 to "straighten" U_2 by an isotopy so that it becomes a linear subspace of $\mathbf{R}^m \times \mathbf{R}^r$ still transversal to $\mathbf{R}^m \times \mathbf{0}$. An obvious isotopy brings it then to coincide with $\mathbf{0} \times \mathbf{R}^r$. These isotopies restricted to the unit disc D^r in U_2 and set to be stationary on M extend by II,5.2 to an isotopy of N that sends $D^r \subset V_2$ to V_1. $\square$

2 Transversality Theorem

The concept of transversality derives its strength from the theorem of Thom asserting that if $f: M \to N$ and V is a submanifold of N, then f can be approximated by maps transversal on V. We will obtain the theorem of Thom as a consequence of the following fundamental theorem:

(2.1) Theorem *Let ξ be a vector bundle over V and let $f: M \to E = E(\xi)$ be a smooth map. Then there is a section $s: V \to E$ such that $f \pitchfork s$.*

Before proving 2.1 we will consider the following situation: We are given a fiber bundle ζ with projection π and base E, and maps $f: M \to E$, $g_1: V \to E(\zeta)$. This yields a diagram

(2.2)

$$M_1 = E(f^*\zeta) \xrightarrow{f_1} E(\zeta) \xleftarrow{g_1} V,$$

$$\pi_1 \downarrow \qquad \pi \downarrow \quad \swarrow g$$

$$M \xrightarrow{f} E$$

where M_1, π_1 are, respectively, the total space and the projection of the induced bundle $f^*\zeta$, f_1 is the natural map, and $g = \pi g_1$. We have:

(2.3) Proposition *If $f_1 \pitchfork g_1$, then $f \pitchfork g$.*

Proof Suppose that $f(p) = g(q)$. We have to show that $Df(T_pM) + Dg(T_qV) = T_{f(p)}E$.

Note first that there is a point p_1 in M_1 such that $f_1(p_1) = g_1(q)$ and $\pi_1(p_1) = p$. The assumption $f_1 \pitchfork g_1$ means that

$$Df_1(T_{p_1}M_1) + Dg_1(T_qV) = T_{f_1(p_1)}E(\zeta).$$

Now apply $D\pi$ to both sides of this and note that $D\pi, D\pi_1$ are both surjective. Thus, by commutativity,

$$T_{f(p)}E = D\pi(T_{f_1(p_1)}E(\zeta)) = D\pi f_1(T_{p_1}M_1) + D\pi g_1(T_qV)$$

$$= Df(T_{\pi_1(p_1)}M) + Dg(T_qV). \qquad \square$$

Proof of 2.1 Assume first that ξ is trivial, i.e., $E = V \times \mathbf{R}^k$. Then 1.3.1 yields a (dense set of points) q in $\mathbf{R}^k$ such that $f \pitchfork V \times \{q\}$. Of course, each such $V \times \{q\}$ can be interpreted as a section of ξ, which proves 2.1 for a trivial bundle ξ.

In the general case there is a bundle η such that $\zeta = \xi \oplus \eta$ is trivial, *cf.* [Bd, II.14.2]. There is a natural projection π of $E(\zeta)$ onto E, which is a projection of a vector bundle. Thus we have the left part of diagram 2.2. Since ζ is trivial, there is a section g_1 transverse to f_1. This completes the diagram 2.2 and we can apply 2.3 to deduce that $g \pitchfork f$. It remains to be shown that g is a section of ξ. This follows from the fact that g_1 is a section and the obvious commutativity of the diagram

where all maps are projections of vector bundles. $\square$

(2.4) Corollary *Let V be a compact submanifold of N, U an open neighborhood of V in N and f: M → N a smooth map. Then there is an isotopy h_t of N that is identity outside of U and such that $f \pitchfork h_1(V)$.*

Proof First, we find a tubular neighborhood of V contained in U and a section s of it transverse to f. By II,5.2, the obvious isotopy of V to $s(V)$ extends to an isotopy of N, which is the identity outside the tubular neighborhood. This completes the proof. $\square$

Compactness can be replaced by the requirement that V be a closed subset of N; instead of applying II,5.2 one constructs the isotopy ad hoc.

(2.5) Corollary *Let $f: M → N$, $g: V → N$ be two maps. Then there is a homotopy h_t of g such that $h_0 = g$ and $h_1 \pitchfork f$.*

Proof Consider the diagram

$$\begin{array}{ccc} M \times V & \xrightarrow{\ f_1\ } & N \times V & \xleftarrow{\ g_1\ } & V, \\ \downarrow & & \downarrow{\scriptstyle \pi} & \nearrow{\scriptstyle g} & \\ M & \xrightarrow{\ f\ } & N & & \end{array}$$

where the vertical maps are projections on the first factor, g_1 is the graph of g, i.e., $g_1(v) = (g(v), v)$, and $f_1(x, v) = (f(x), v)$. Then, $V_1 = g_1(V)$ is a submanifold of $N \times V$ and by 2.1 there is a section $s: V_1 → N \times V$ of its tubular neighborhood transverse to f_1. Observe now that $f_1 \pitchfork sg_1$ and let H_t be an isotopy of $g_1(V)$ to $sg_1(V)$. Then $\pi H_t = h_t$ is a homotopy of g to a map h_1, which by 2.3 is transverse to f. $\square$

3 Morse Functions

Suppose now that we are given a real valued function $f: M \to \mathbf{R}$. If, at a point $p \in M$, Df is non-degenerate, then, as we know, f at p is equivalent to a projection: non-degenerate in this case means the same as being of maximal rank. If Df is degenerate at p, i.e., p is a critical point, then the local behavior of f at p can be quite complicated. A fundamental idea due to M. Morse was to single out a class of functions with a particularly nice behavior at critical points and to show that they form a dense set. "Nice behavior" means that at critical points they behave like—i.e., are equivalent to—one of the quadratic functions $\sum \delta_i x_i^2$ at 0, $\delta_i = \pm 1$. In particular, the list of possibilities is—up to equivalence—finite.

As usual, we prefer an invariant definition and the easiest way is to work in the cotangent space. Recall that, given $f: M \to \mathbf{R}$, $df: M \to T^*M$ is the section of the cotangent bundle given at $p \in M$ by $df(X) = X(f)$, $X \in T_pM$.

(3.1) Definition We say that $p \in M$ is *critical* if $df = 0$ at p, i.e., if df intersects the zero section M_0 of the cotangent bundle above p. We say that p is a *non-degenerate critical point* if this intersection is transversal. A function f which has only nondegenerate critical points, that is, such that $df \pitchfork M_0$, is called a *Morse function*.

It follows immediately from 1.5 that:

(3.2) Lemma *Critical points of a Morse function are isolated.*

We will delay for a moment the investigation of the local behavior of Morse functions and begin by showing that there are, indeed, a lot of them.

(3.3) Lemma *Let M be a submanifold of $\mathbf{R}^k$ and let $f: M \to \mathbf{R}$. There is a dense set of linear functions $L: \mathbf{R}^k \to \mathbf{R}$ such that $f - L$ restricted to M is a Morse function.*

Proof We will build a diagram of spaces and maps in the following way: Begin with the cotangent bundle of $\mathbf{R}^k$ restricted to M, i.e., $T^*\mathbf{R}^k | M$. This is also a bundle over T^*M with the projection π. Then the map $df: M \to T^*M$ yields the induced bundle with total space E and all this forms the commutative square on the left in the diagram:

$$E \xrightarrow{\; g \;} T^*\mathbf{R}^k \,|\, M \xleftarrow{\; dL|M \;} M.$$

To get the triangle on the right, note that $T^*\mathbf{R}^k \,|\, M$ is a trivial bundle, hence by 1.3.1 there is a dense set of constant sections $M \times \{q\}$ that are transverse to g. A constant section is a differential of a linear map $L: \mathbf{R}^k \to \mathbf{R}$. Thus to complete the diagram we choose as L a linear map such that $dL|M \pitchfork g$ and observe that $\pi \circ dL \,|\, M = d(L|M)$.

Now, 2.3 implies that $df \pitchfork d(L|M)$, i.e., that $d(f - L|M)$ is transversal to the zero section. $\square$

(3.4) Theorem *Given $f: M \to \mathbf{R}$ and $\varepsilon > 0$, there is a Morse function $g: M \to \mathbf{R}$ such that $|f - g| < \varepsilon$.*

Proof Consider M as a submanifold of the unit ball in an $\mathbf{R}^k$ and take as L in 3.3 a linear function such that $|L| < \varepsilon$ in the ball. $\square$

Now, let M be a manifold with compact boundary and suppose that $\partial M = V_0 \cup V_1$ where the V_i are disjoint and compact.

(3.5) Theorem *There is a Morse function $f: M \to I$ such that*:

(a) *f has no critical points in a neighborhood of ∂M*;
(b) *$f^{-1}(i) = V_i$, $i = 0, 1$.*

Proof Let $\partial M \times [0, 1) \subset M$ be a collar of ∂M. By I,7.4 there is a smooth function $g: M \to I$ with the following properties:

$$g(x, t) = t \quad \text{for } (x, t) \in V_0 \times [0, \tfrac{1}{2}],$$

$$g(x, t) = 1 - t \quad \text{for } (x, t) \in V_1 \times [0, \tfrac{1}{2}],$$

$$1/4 < g(x) < 3/4 \quad \text{elsewhere.}$$

Then g has properties (a) and (b) but is not necessarily Morse. To obtain a Morse function we assume that M is a submanifold of the unit ball in an $\mathbf{R}^k$ and consider the function $f = g + \mu L$, where $\mu: M \to I$ is smooth, equals 0 in $\partial M \times [0, 1/4]$ and equals 1 in $M - \partial M \times [0, \frac{1}{2}]$, and L is a still to be chosen linear map of $\mathbf{R}^k$.

Clearly, f satisfies (a) and, if $|L| < 1/4$ in M, then it satisfies (b) as well. Assume that some Riemannian metric is given in T^*M.

Since $|d(\mu L)| \le |d\mu| |L| + \mu |dL|$ we see that by taking L "small" we can make $|d(\mu L)|$ as small as we want in the compact set $\partial M \times [0, 1/2]$. In particular, since $|dg|$ is bounded away from 0 in this set, we can achieve that

$$|d(g + \mu L)| \ge |dg| - |d(\mu L)| > 0 \qquad \text{in } \partial M \times [0, 1/2],$$

i.e., that f has no critical points there. Then, if L is such that $g + L$ is Morse in M, the same is true of $f = g + \mu L$. $\square$

It is sometimes convenient to require that the function f in 3.5 has the following additional property:

(3.5)   (c) *f takes distinct values at distinct critical points.*

This is easily achieved as follows: If x is a critical point of f then, by 3.2, there is a pair of neighborhoods U, V of x such that $\text{Cl}(U) \subset V$, $Cl(V)$ is compact, and x is the only critical point of f in V. Let $\mu: M \to I$ equal 1 in U and 0 outside of V. Then, for small c, $f + c\mu$ has the same critical points as f, but the critical value at x is changed by c. The argument is similar, but simpler, to that used in the proof of 3.5 and is left as an exercise.

4   Neighborhood of a Critical Point

There remains to investigate the behavior of a Morse function in a neighborhood of a critical point.

Suppose that p is a critical point of $f: M \to \mathbf{R}$ and choose a local chart at p. The Hessian of f at p is the matrix of second derivatives of f at p. It depends on the choice of the local chart. However:

(4.1)   Lemma *Let p be a critical point of f. Then p is non-degenerate if and only if the Hessian of f at p is of maximal rank.*

Proof A choice of a chart in a neighborhood U of p also gives a trivialization of the cotangent bundle restricted to U, that is, a projection $\phi: T^*M \,|\, U \to T_p^*M$. p is non-degenerate if and only if $0 \in T_p^*M$ is a regular value of $\phi\, df$, i.e., if the differential of this map at p is surjective. In the chosen local coordinate system this means that the Jacobian of $\phi\, df$ is to be of maximal rank. However, the map $\phi\, df$ simply assigns to every point the coordinates of df at this point; thus its Jacobian is the Hessian of f at p. □

(4.2) Proposition *Suppose that p is a non-degenerate critical point of f. Then in some system of local coordinates at p, f is given by $f(p) + \sum_i \delta_i x_i^2$, $\delta_i = \pm 1$.*

Proof Let f be a real valued function defined in a neighborhood of $0 \in \mathbf{R}^m$. Suppose that the Hessian of f at 0 is of maximal rank and that $f(0) = 0$. We have to show that there is a diffeomorphism h of a neighborhood of 0 such that

$$fh(x_1, x_2, \ldots, x_m) = \sum_{i \le k} x_i^2 - \sum_{i > k} x_i^2.$$

This will be done in two steps. In the first we show that

$$(*) \qquad\qquad f(x) = \sum_{i,j} h_{ij} x_i x_j,$$

where the h_{ij} are some functions of x and $h_{ij} = h_{ji}$. Thus f looks like a symmetric bilinear form—but with variable coefficients—which suggests that we should try to adapt one of usual procedures of diagonalization of such forms to our situation. This works, and that is the second step of the proof. Now the details.

Since f has a critical point at 0 we have, by A,2.2,

$$f(x) = \sum_i h_i(x) x_i,$$

where $h_i(0) = (\partial f/\partial x_i)(0) = 0$. We can apply the same lemma once more to h_i to get $h_i = \sum_j h_{ij} x_j$. Now, setting $h_{ij} = \frac{1}{2}(h_{ij} + h_{ji})$ we finally obtain $(*)$.

The diagonalization of f is now done inductively. Suppose that in some chart f is already in the form

$$f(x) = \pm x_1^2 + \cdots \pm x_{k-1}^2 + \sum_{i,j \ge k} h_{ij} x_i x_j, \qquad h_{ij} = h_{ji}.$$

Through a linear change of coordinates we can achieve that $h_{kk}(0) \ne 0$; hence $h_{kk}(x) \ne 0$ in a certain neighborhood U of 0. Consider the transformation $F: U \to \mathbf{R}^m$ given by

$$y_i = x_i \qquad \text{for } i \neq k,$$

$$y_k = |h_{kk}|^{1/2}\left(x_k + \sum_{i>k} \frac{h_{ik}x_i}{|h_{kk}|}\right).$$

The Jacobian of F at $\mathbf{0}$ does not vanish: Its determinant equals $|h_{kk}(\mathbf{0})|^{1/2}$. Therefore F is a diffeomorphism in a neighborhood $V \subset U$ of $\mathbf{0}$ in $\mathbf{R}^m$. Since

$$fF^{-1}(y) = \sum_{i \leq k} \pm y_i^2 - \sum_{i,j>k} \frac{h_{ij}h_{jk}}{h_{kk}} y_i y_j,$$

this concludes the inductive step. $\square$

The number of minus signs in this local representation of f at a critical non-degenerate point p does not depend on the choice of chart; it is called the *index* of p. To see this let $t = f(p)$ and let $M_t = \{q \in M \,|\, f(q) \leq t\}$. Suppose that in some local chart f is given by $-\sum_{i \leq k} x_i^2 + \sum_{i > k} x_i^2$ and let $T = \{x \in \mathbf{R}^n \,|\, \sum_{i>k} x_i^2 \leq \sum_{i \leq k} x_i^2\}$. Then T is a cone on $S^{k-1} \times D^{m-k}$ with the vertex at $\mathbf{0}$; hence $H_*(T, T - \mathbf{0}) \simeq H_*(\mathbf{R}^k, \mathbf{R}^k - \mathbf{0})$. By an obvious excision argument $H_*(M_t, M_t - p) \simeq H_*(T, T - \mathbf{0})$, which shows that k can be read from the local homology properties.

5   Intersection Numbers

Using the notion of transversality, we will define here intersection numbers.

Let V^r, M^m be compact closed transversal submanifolds of N^{m+r} and let $V \cap M = \{p_1, \ldots, p_k\}$. It follows from 1.7 that there is a tubular neighborhood F of M such that fibers F_{p_i} of F at p_i are open neighborhoods of the p_i in V. Therefore, if both V and the normal bundle of M are oriented, we can compare the induced local orientation at every point p_i; we set $\varepsilon(p_i) = +1$ if the orientation of F_{p_i} agrees with the orientation of V; otherwise $\varepsilon(p_i) = -1$, cf. I,3.6. Finally, let

$$[V : M] = \sum_i \varepsilon(p_i);$$

this is the *intersection number* of V and M.

To explain the significance of this number recall that if E is the total space of an oriented r-dimensional bundle over a connected manifold M then for every fiber E_p the inclusion $j: E_p \hookrightarrow E$ induces an isomorphism $H_r(E_p, E_p - \mathbf{0}) \to H_r(E, E - E_0)$.

(This is a special case of the so-called Thom isomorphism, *cf.* [Sp, p. 259]. It can be proved directly by noticing that it is true for a trivial bundle and then using the Mayer–Vietoris theorem.)

For the case under consideration this means that the orientation of the normal bundle to a connected manifold M produces a well-defined generator of $H_r(F, F - M)$, hence also of $H_r(N, N - M)$; we denote the latter γ_M.

Now, if $\gamma_i \in H_r(V, V - p_i)$ is a local orientation of V at p_i, then the inclusion $(V, V - p_i) \hookrightarrow (N, N - M)$ sends γ_i to $\varepsilon(p_i)\gamma_M$. Therefore

$$(5.1) \qquad j_*\gamma_V = [V:M]\gamma_M,$$

where $\gamma_V \in H_r(V)$ is the orientation of V and j_* the composition $H_r(V) \to H_r(V, V - \bigcup_i p_i) \to H_r(N, N - M)$. Since j_* can also be expressed as the composition $H_r(V) \to H_r(N) \to H_r(N, N - M)$, this shows that the intersection number does not depend on the isotopy class of the imbedding $V \subset N$. Therefore it follows from 2.4 and 5.1 that we do not have to assume the transversality to define it.

To define the intersection number of two oriented submanifolds V, M of an oriented manifold N we have to agree how these data determine the orientation of the normal bundle to M. We accept the convention that at every point $p \in M$ the orientation of M followed by the orientation of the fiber of its tubular neighborhood agrees with the given orientation of N at p. With this convention in force $[V:M]$ can be computed directly by noticing that, at every point $p \in V \cap M$, $\varepsilon(p) = +1$ if the orientation of M at p followed by the orientation of V at p agrees with the orientation of N at p. (This makes sense by 1.6.) It follows that

$$(5.2) \qquad [V:M] = (-1)^{rm}[M:V].$$

We will now derive a formula expressing the intersection number of two cross sections of a k-disc bundle over S^k in terms of the characteristic element of the bundle. For this we first establish a lemma expressing the degree of a map as an intersection number.

Let $f: (M, \partial M) \to (N, \partial N)$ be a smooth map of oriented compact connected manifolds of the same dimension k. The degree of f is then defined as the integer d_f satisfying

$$f_*\gamma_M = d_f\gamma_N,$$

where γ_M, γ_N are respective orientations, i.e., generators of $H_k(M, \partial M)$ and $H_k(N, \partial N)$. Let $g: M \to M \times N$ be the graph of f, $\pi: M \times N \to N$ the projection, and $M_p = M \times \{p\}$, $p \in \text{Int } N$. Consider the diagram

$$H_k(M, \partial M) \xrightarrow{\ g_* \ } H_k(g(M), g(\partial M)) \xrightarrow{\ j_* \ } H_k(M \times N, M \times N - M_p)$$

$$\begin{array}{ccc} & \searrow{\scriptstyle f_*} & \downarrow{\scriptstyle \pi_*} & \qquad \qquad \pi_* \downarrow \simeq \\ & & H_k(N, \partial N) \xrightarrow{\ \simeq \ } H_k(N, N - p). \end{array}$$

We have, by 5.1, $j_* g_* \gamma_M = [g(M): M_p]\gamma$, where γ is the generator sent by π_* to the local orientation of N at p and νM_p is oriented by γ. Since the diagram is commutative, we have:

(5.3) Lemma $d_f = [g(M): M_p]$. $\qquad \square$

The sign in in this formula remains unchanged if we view $g(M)$ and M_p as oriented by the projection on M, and $M \times N$ with the product orientation. The geometric content is particularly clear if p is chosen to be a regular value of f (see Fig. IV,2).

Let now S_1 and S_2 be two cross sections of a k-disc bundle over S^k with characteristic element $\alpha \in \pi_{k-1}(\mathbf{SO}(k))$.

(5.4) Proposition $[S_1: S_2] = \phi_*(\alpha)$.

Here ϕ is the projection of the bundle $\mathbf{SO}(k)/\mathbf{SO}(k-1) = S^{k-1}$ and we have identified $\pi_{k-1}(S^{k-1})$ with $\mathbf{Z}$; cf. A,5 for all relevant notions.

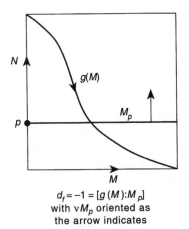

$d_f = -1 = [g(M):M_p]$
with νM_p oriented as
the arrow indicates

Figure IV,2

Proof Let D, D_1, D_2 be three copies of the k-disc D^k. Then the bundle in question is obtained from the disjoint union $D_1 \times D \cup D_2 \times D$ by identifying $\partial D_1 \times D$ with $\partial D_2 \times D$ via the map $(x, y) \mapsto (x, \alpha(x) \cdot y)$. We can assume using appropriate isotopies that the section S_1 is the zero section of the bundle and that S_2 over D_2 is the constant cross section $(x, \mathbf{e})$, where $\mathbf{e}$ is the first basis vector of $\mathbf{R}^k$. This means that over D_1 the section S_2 is the graph of a map $s: (D_1, \partial D_1) \to (D, \partial D)$ such that $s|\partial D_1$ is given by $x \mapsto \alpha(x) \cdot \mathbf{e}$, i.e., $s|\partial D_1 = \phi\alpha$.

Now, all intersections of S_1 and S_2 are over D_1: thus 5.3 yields $d_s = [S_1 : S_2]$. Since $d_s = d_{s|\partial D_1} = d_{\phi\alpha} = \phi_*(\alpha)$, 5.4 follows. $\square$

In particular, letting S be the zero section we obtain

(5.4.1) $$[S : S] = \phi_*(\alpha).$$

Exercise Consider a smooth map $f: S^k \times S^k \to S^k$. Let $S_1 = S^k \times \{a\}$, $S_2 = \{a\} \times S^k$, $f_i = f | S_i$, and let d_i be the degree of f_i, $i = 1, 2$.
   Show that $d_1 = [f^{-1}(b) : S_2]$, $b \in S^k$.
   The pair (d_1, d_2) is called the *bidegree* of f.

6   Historical Remarks

The concept of transversality has its roots in the notion of general position studied extensively in the piecewise linear context. It was introduced into differential topology by R. Thom in 1954 in [T2]. It was Proposition 1.4 that by associating *manifolds* to *maps* allowed Thom to construct the cobordism theory. Theorem 2.1 is stated in that paper as an approximation theorem. A far-reaching generalization of it, needed in the study of singularities of differentiable maps, was found by Thom in [T3].

The notion of critical and critical non-degenerate point, appears for the first time in a seminal paper of M. Morse in 1925 [Mo1]. This paper contains 4.2 (with the same proof as here) but not the approximation theorem 3.4, which appeared only in 1934 in [Mo2].

One of the important directions of research generated by 4.2 is the theory of singularities of smooth maps. The problem is, *grosso modo*, to describe a class of maps that have singularities from a given list only, and which form a dense subset. This program has been carried out by H. Whitney in the case of maps $\mathbf{R}^2 \to \mathbf{R}^2$ [Wi5]; the general case was studied by Thom in [T3].

Morse functions were initially utilized to establish certain relations between the number of critical points of various types and the homology of manifold. (We prove these relations—Morse inequalities—in Chapter VII.) The underlying idea was to investigate the change in the topological character of manifolds M_t as t passes through a critical value. This idea, which can be traced back to Poincaré [P3, §2], led eventually to the handle presentation of a manifold, which we will study in Chapters VII and VIII.

V

Foliations

A subbundle of the tangent bundle of a manifold M can be thought of as a field of hyperplanes. Some such fields arise naturally in a geometric context. For instance, a projection $\pi: M \to N$ of a fiber bundle gives rise to field of planes tangent to fibers, i.e., the field Ker $D\pi$. In this chapter we investigate relations between subbundles of the tangent bundle and algebraic structures derived from the fact that the set $\mathscr{X}(M)$ of vector fields on M is a vector space, even a Lie algebra, as well as a module over the ring of smooth functions on M. Thus, in Section 1, the module structure is utilized to show that subbundles of the tangent bundle are in one to one correspondence with a certain class of submodules. In Section 2 we introduce the concept of foliation. It generalizes that of a fibration and, like it, gives rise to a subbundle of the tangent bundle. The subbundles obtained in this way are characterized in terms of the Lie algebra structure on $\mathscr{X}(M)$ in Section 3. In Section 4 we prove the most prominent geometric property of foliations, the existence of *leaves*. A few examples are collected in the last section.

This chapter contains only a very elementary introduction to a beautiful geometric subject. An extensive treatment can be found in the book of C. Godbillon [Go].

1 *d*-Fields

Recall that the set $\mathscr{X}(M)$ of vector fields on a manifold M is a module over the ring $C^\infty(M)$ of smooth functions on M (*cf.* I,5.2).

(1.1) *Definition* A family $\{X^\alpha\}$ of vector fields is *locally finite* if every $p \in M$ has a neighborhood in which almost all fields X^α vanish. A submodule V of $\mathscr{X}(M)$ is *complete* if for every locally finite family X^α of vector fields in V their sum $\sum_\alpha X^\alpha$ is in V. (Almost all = all but a finite number.)

If M is non-compact, then the module V of vector fields that have the property that each vanishes outside of a compact set is not complete. For let X be a vector field not in V and let $\{\lambda_\alpha\}$ be a partition of unity associated to an adequate atlas. Then the family $\{\lambda_\alpha X\}$ is locally finite and each field $\lambda_\alpha X$ is in V, but $\sum_\alpha \lambda_\alpha X = X \notin V$.

Exercise If M is compact, then every submodule of $\mathscr{X}(M)$ is complete.

If V is a subspace of $\mathscr{X}(M)$ and $p \in M$, then we denote by V_p the set of vectors of T_pM that belong to fields from V. Clearly, V_p is a subspace of T_pM. We say that V is of dimension d if, for all $p \in M$, V_p is of dimension d. (This is not the same as the dimension of V as a vector space. The last one is infinite in all cases of interest here.)

We are now ready to state the main theorem of this section. For brevity, we call a d-dimensional smooth subbundle of TM a *d-field*. If E is a d-field, then $V(E)$ denotes the set of all vector fields that lie in E.

(1.2) **Theorem** *Let E be a d-field on M. Then $V(E)$ is a d-dimensional complete submodule of $\mathscr{X}(M)$, and this construction establishes a one-to-one correspondence between d-fields on M and complete d-dimensional submodules of $\mathscr{X}(M)$.*

Proof It is trivial that $V(E)$ is a submodule; to show that it is complete suppose that $\{X^\alpha\}$ is a locally finite family of vector fields from $V(E)$ and let $X = \sum_\alpha X^\alpha$. Then $X_p = \sum_\alpha X^\alpha_p$, and the sum on the right is a sum of a finite number of vectors in E_p, hence also a vector in E_p. Thus $X_p \in E_p$, i.e., $X \in V(E)$.

To show that $V(E)$ is of dimension d it is enough to show that $V(E)_p = E_p$. Clearly, $V_p \subset E_p$. The reverse inclusion will follow if we show that through every vector $e \in E_p$ passes a vector field that lies in E. To see this, let $X^1, \ldots, X^d$ be vector fields on M such that $X^1, \ldots, X^d$ span E_q at all points q in a neighborhood U of p. Then $e = \sum_i \mu_i X^i_p$ for some μ_i. Let λ be a smooth function on M which equals 1 at p and vanishes outside of a compact subset of U. Set

$$X_q = \lambda(q) \sum_i \mu_i X^i_q;$$

this defines a vector field in $V(E)$ and such that $X_p = e$.

Now let V be a d-dimensional submodule and let $E(V) = \bigcup_{p \in M} V_p$. We claim that $E(V)$ is a d-field. For let $X^1, \ldots, X^d$ be vector fields in V such that $X^1_p, \ldots, X^d_p$ span V_p. Then there exists a neighborhood U of p such that $X^1_q, \ldots, X^d_q$ are linearly independent at all $q \in U$. Therefore they span V_q for all $q \in U$. This gives smooth local product structure in $E(V)$ showing that $E(V)$ is a d-field.

Now, beginning with a d-field E and forming successively $V(E)$ and $E(V(E))$, we have $E(V(E)) = \bigcup_{p \in M} V(E)_p = \bigcup_{p \in M} E_p = E$, i.e., the map $E \mapsto V(E)$ has a left inverse. To show that it has a right inverse we have to show that $V(E(V)) = V$. Certainly $V(E(V)) \supset V$. To establish the reverse inclusion we have to show that if a vector field X satisfies $X_p \in V_p$ for all $p \in M$, then X is in V.

We have just seen that there is a covering $\{U_\alpha\}$ of M and, for every α, d vector fields $X^{1,\alpha}, \ldots, X^{d,\alpha}$ in U_α that span V_q for every $q \in U_\alpha$. Therefore, in U_α, $X_q = \sum_i \mu_i(q) X^{i,\alpha}_q$, where the μ_i are smooth functions in U_α. We can assume that $\{U_\alpha\}$ is an adequate covering. Let $\{\phi^\alpha\}$ be an associated partition of unity, and consider the vector field X^α defined on M by

$$X^\alpha = \phi^\alpha \sum_i \mu_i X^{i,\alpha}.$$

Then X^α is a vector field in V and it vanishes outside of U_α. Thus the family $\{X^\alpha\}$ is locally finite and, since V is complete, $\sum_\alpha X^\alpha$ is in V. But $\sum_\alpha X^\alpha = X$. □

Exercise Let $m = \dim M$. Show that $\mathscr{X}(M)$ is generated (as a module over $C^\infty M$) by m vector fields if and only if TM is a trivial bundle.

Exercise Let M be the Möbius band viewed as a bundle over the circle. The bundle tangent to fibers (*cf.* III,1) is a 1-field such that the corresponding 1-dimensional submodule of $\mathscr{X}(M)$ is not generated by a vector field.

2 Foliations

A submersion $M^m \to N^{m-d}$ gives rise to a d-field on M, but to have a d-field it is not necessary for the submersion to be defined globally: it is enough to have a "field" of local submersions satisfying a natural compatibility condition. This suggests the notion of an atlas of submersions:

(2.1) *Definition* An *atlas of submersions* of codimension d on M consists of a covering $\{U_\alpha\}$ of M and a family $\{f_\alpha\}$ of submersions $f_\alpha\colon U_\alpha \to \mathbf{R}^{m-d}$ of rank $m - d$ satisfying the following compatibility condition: If $p \in U_\alpha \cap U_\beta$, then there is a diffeomorphism (into) $h\colon U \to \mathbf{R}^{m-d}$ of a neighborhood U of $f_\alpha(p)$ such that $f_\beta = hf_\alpha$ in a neighborhood of p.

Two atlases are compatible if their union is an atlas. As it was in the case of differential structures, every atlas of submersions determines uniquely a maximal atlas: the union of atlases that contain it. We now conclude definition 2.1:

A maximal atlas of submersions of codimension d is called a *foliation of codimension d*, or, simply, a *d-foliation*.

To every d-foliation $\mathcal{F}$ on M we can associate a d-field $E(\mathcal{F})$ as follows: If $p \in U_\alpha \cap U_\beta$, then the compatibility relation guarantees that $\operatorname{Ker} D_p f_\alpha = \operatorname{Ker} D_p f_\beta$. Therefore in every tangent space $T_p M$ there is a well-defined d-dimensional subspace E_p. Let $E(\mathcal{F}) = \bigcup_{p \in M} E_p$.

(2.2) Theorem $E(\mathcal{F})$ *is a d-field. Moreover, if $E(\mathcal{F}_1) = E(\mathcal{F}_2)$, then $\mathcal{F}_1 = \mathcal{F}_2$.*

Proof Let $\{U_\alpha, f_\alpha\}$ be an atlas of submersions for $\mathcal{F}$, and let $p \in U_\alpha$. By II,1.2 there is a system of local coordinates $x_1, \ldots, x_m$ in a neighborhood U of p such that $f_\alpha(x_1, \ldots, x_m) = (x_1, \ldots, x_{m-d})$. Then, at all points of U, $\operatorname{Ker} Df_\alpha$ is spanned by $\partial_{m-d+1}, \ldots, \partial_m$. This shows that $E(\mathcal{F})$ is a d-field.

The second part of the theorem is a consequence of the following:

(2.3) Lemma *Let U be an open subset of $\mathbf{R}^m$ and let $f, g\colon U \to \mathbf{R}^{m-d}$ be two submersions such that $\operatorname{Ker} Df = \operatorname{Ker} Dg$. Given $p \in U$, there is a neighborhood W of $f(p)$ and a diffeomorphism (into) $h\colon W \to \mathbf{R}^{m-d}$ such that $hf = g$ in a neighborhood of p.*

Proof Let $\pi: \mathbf{R}^m \to \mathbf{R}^{m-d}$ be the standard projection. We have shown that $f(x_1, \ldots, x_m) = \pi(x_1, \ldots, x_m) = (x_1, \ldots, x_{m-d})$ in a system of local coordinates $x_1, \ldots, x_m$ in a neighborhood U of p. Since $\operatorname{Ker} Dg = \operatorname{Ker} Df$ and $\operatorname{Ker} Df$ is spanned by $\partial_{m-d+1}, \ldots, \partial_m$ in U, we have $Dg(\partial_{m-d+j}) = 0, j = 1, \ldots, d$. This means that $g \mid U$ does not depend on the last d coordinates $x_{m-d+1}, \ldots, x_m$; that is, it factors as $g = h\pi$, where h is defined in a neighborhood $\pi(U)$ of $\mathbf{0} \in \mathbf{R}^{m-d}$. It is easily seen that the rank of h is $m - d$; hence, by the Implicit Function Theorem (*cf.* A,1.1), there is a neighborhood W of $\mathbf{0}$ such that $h \mid W$ is a diffeomorphism $W \to h(W)$. Now, $hf = h\pi = g$ in $U \cap f^{-1}W$. $\square$

Another useful consequence of 2.3 is the following:

(2.4) Corollary *Let E be a d-field on M and suppose that there is a covering $\{U_\alpha\}$ of M and submersions $f_\alpha: U_\alpha \to \mathbf{R}^{m-d}$ such that $\operatorname{Ker} D_q f_\alpha = E_q$, $q \in U_\alpha$. Then $\{U_\alpha, f_\alpha\}$ is an atlas of submersions.*

Proof By 2.3, the f_α satisfy compatibility relations. $\square$

A d-field E such that $E = E(\mathscr{F})$ for some foliation $\mathscr{F}$ is called *completely integrable*. Corollary 2.4 asserts that this is a local property of d-fields; we will study it in the next section. An important example of completely integrable fields is given by the following:

(2.5) Proposition *Every line field is completely integrable.*

Proof By 2.4, it is enough to consider a neighborhood of a point $p \in M$ in which the line field is spanned by a nonvanishing vector field. Assume then that X is a nonvanishing vector field in a neighborhood U of $\mathbf{0} \in \mathbf{R}^m$ with $X_0 = \partial_m$; it follows from I,6.1. that there is a neighborhood W of $\mathbf{0}$ and a map $f: W \times (-\varepsilon, \varepsilon) \to U$, where ε is a positive number, such that $f(p, 0) = p$ and $Df(\partial t) = X$.

Consider now the manifold $V' = (W \cap \mathbf{R}^{m-1}) \times (-\varepsilon, \varepsilon)$; let $h = f \mid V'$ and let $\pi: V' \to \mathbf{R}^{m-1}$ be the natural projection. Then $Dh_0\partial_i = \partial_i$ for $i = 1, \ldots, m - 1$ and $Dh_0 \partial t = X_0$, that is, h is of maximal rank at $\mathbf{0}$. It follows that there is a neighborhood V of $\mathbf{0}$ in V' on which h is a diffeomorphism. Hence $\pi h^{-1}: h(V) \to \mathbf{R}^{m-1}$ is a submersion such that the kernel of its differential is generated by X. $\square$

Observe that in the course of this proof we established that:

(2.6)   *If X is a nonvanishing vector field in a neighborhood U of $\mathbf{0} \in \mathbf{R}^m$, then there is a neighborhood $V \subset U$ and a diffeomorphism $h: V \to U$ such that $Dh(\partial_m) = X$.*

3   Frobenius Theorem

Not every d-field is completely integrable. To characterize those which are, we need the notion of the bracket $[X, Y]$ of vector fields defined in I,5 by the equation

$$[X, Y] = XY - YX.$$

As an example, observe that in $T\mathbf{R}^m$ we have $[\partial_i, \partial_j] = 0$ by the well-known theorem of elementary Calculus.

The following proposition is verified by routine calculation.

(3.1)   Proposition

(a) $[X, Y]$ *is linear in each factor and skew-symmetric*:
  $[X, Y] = -[Y, X]$;
(b) $[[X, Y], Z] + [[Y, Z], X] + [[Z, X], Y] = 0$;
(c) $[fX, gY] = fg[X, Y] + fX(g)Y - gY(f)X.$ $\square$

A vector space with an operation $[\ ,\ ]$ satisfying (a) and (b) is called a Lie algebra. Property (c) shows that it makes sense to speak of submodules of $\mathscr{X}(M)$ that are also Lie algebras, that is, are closed under the bracket operation.

Exercise   Show that a 1-dimensional submodule of $\mathscr{X}(M)$ is a Lie algebra.

Submodules that are Lie algebras arise naturally in the context of foliations. For suppose that X and Y are two vector fields on M annihilated by a map $f: M \to N$ (i.e., $(Df)X = 0 = (Df)Y$). Then $(Df)[X, Y] = 0$, for $((Df)[X, Y])g = X(Y(gf)) - Y(X(gf)) = 0$ for every smooth function g on N. This implies that if E is a completely integrable d-field on M and X, Y lie in E (i.e., $X, Y \in V(E)$), then $[X, Y] \in V(E)$. In other words, $V(E)$ is a Lie algebra. It turns out that this property characterizes completely integrable fields. This is content of the following theorem of Frobenius:

(3.2) Theorem *Let E be a d-field on M. Then E is completely integrable if and only if $V(E)$ is a subalgebra of $\mathcal{X}(M)$.*

Proof We have just seen that the condition is necessary. Moreover, it follows from 2.5 and the preceding exercise that the theorem is true for $d = 1$. We will therefore assume it to be true for $(d - 1)$-dimensional fields, $d \geq 2$, and prove that a d-dimensional field E such that $V(E)$ is a Lie algebra is completely integrable.

Let U be a chart in M and let $X^1, \ldots, X^d$ be d vector fields in U such that $X^1_q, \ldots, X^d_q$ span E_q at every $q \in U$. We view U as a neighborhood of $\mathbf{0}$ in $\mathbf{R}^m$ and assume that

$$(*) \qquad X^1 = \partial_1, \qquad X^i = \sum_{j=2}^{m} a^i_j \partial_j \quad \text{for } i \geq 2.$$

The legitimacy of the first assumption follows from 2.6; to ensure the second we may have to replace X^i, $i \geq 2$, by $X^i - X^i(x_1)\,\partial_1$, i.e., to subtract the projection on $X^1 = \partial_1$.

Let $S = \{(x_1, \ldots, x_m) \in U \mid x_1 = 0\}$ and let $Y^2, \ldots, Y^d$ be the vector fields $X^2, \ldots, X^d$ restricted to S. By $(*)$ they are linearly independent and tangent to S; hence they span a $(d - 1)$-field E_S on S. Since $[\partial_i, \partial_j] = 0$, it follows from 3.1(c) that $[Y^i, Y^j]$, $i, j > 1$, is a linear combination of $Y^2, \ldots, Y^d$, i.e., $V(E_S)$ is a Lie algebra. We now apply the inductive assumption and deduce that there is a submersion $f: S' \to \mathbf{R}^{m-d}$, where S' is a neighborhood of $\mathbf{0}$ in S, such that $\operatorname{Ker} Df = E_S$. We will complete the proof of the theorem by showing that $f\pi: \pi^{-1}(S') \to \mathbf{R}^{m-d}$ is the submersion we have been looking for, that is, that $\operatorname{Ker} Df\pi = E$, $\pi(x_1, \ldots, x_m) = (0, x_2, \ldots, x_m)$.

Certainly, $f\pi$ is a submersion and $(Df\pi)X^1 = 0$. Let $f = (f_1, \ldots, f_{m-d})$, $f_j = f_j(x_2, \ldots, x_m)$. Then

$$(Df\pi)X^i = (X^if_1, \ldots, X^if_{m-d}).$$

To show that $X^if_j = 0$, we note first that $X^1f_j = 0$ for all j; thus $X^1X^if_j = X^1X^if_j - X^iX^1f_j = [X^1, X^i]f_j = \sum_{k=1}^{d} c_{ik}X^kf_j = \sum_{k=2}^{d} c_{ik}X^kf_j$.

Letting $X^if_j = \beta^i_j$, $\beta^i_j = \beta^i_j(x_1, \ldots, x_m)$, this can be rewritten as

$$(**) \qquad \frac{\partial}{\partial x_1}\beta^i_j = \sum_{k=2}^{d} c_{ik}\beta^k_j, \qquad i = 2, \ldots, d, \ j = 1, \ldots, m - d.$$

For a fixed index j and a point $q = (x_2, \ldots, x_m) \in S'$ the system $(**)$ becomes a homogeneous system of $d - 1$ differential equations for $d - 1$

functions $\beta_j^i(x_1, q)$ of the variable x_1. The initial condition at $x_1 = 0$ is

$$\beta_j^i(0, q) = X^i f_j \qquad \text{at } q \in S'$$

$$= Y^i f_j$$

$$= 0, \qquad \text{since } Y^i \in \text{Ker } Df.$$

By the Uniqueness Theorem for solutions of such a system, $\beta_j^i(x_1, q) = 0$ for all x_1, that is, $X^i f_j = 0$ for all i, j. This shows that $E \subset \text{Ker } Df\pi$. Since $f\pi$ is a submersion, $E = \text{Ker } Df\pi$ for dimensional reasons. $\qquad \square$

Collecting together 1.2, 2.2, and 3.2 we obtain the following:

(3.3) Corollary *There is a one-to-one correspondence between the set of d-foliations on M, the set of completely integrable d-fields, and the set of d-dimensional complete submodules of $\mathscr{X}(M)$ that are also subalgebras.* $\qquad \square$

Frobenius's theorem, 3.2, is local in character and the condition for complete integrability is not topological. It is reasonable to ask the following global question: Given a subbundle E of TM, does there exist an isomorphic subbundle that is completely integrable? R. Bott has given a necessary condition for this to be true in terms of characteristic classes of E. An exposition of these results is in [B1]. This book also contains a clear presentation of A. Haefliger's theory of classifying spaces for foliations.

4   Leaves of a Foliation

In the last section we developed the local theory of foliations. An important global concept is that of a leaf. It generalizes the notion of fiber of a fiber bundle: A fiber bundle is a union of fibers; a foliated space is a union of leaves.

(4.1) Definition Let $\mathscr{F}$ be a foliation on M. An *integral manifold* of $\mathscr{F}$ is a pair (N, f) where N is a manifold and $f: N \to M$ is a one-to-one immersion such that $(Df)T_q N \subset E_{f(q)}(\mathscr{F})$.

A *leaf* of $\mathscr{F}$ is a maximal connected integral manifold: that is, an integral manifold (N, f) is a leaf if N is connected and whenever a connected integral manifold (N_1, f_1) intersects it, there is a one-to-one immersion $g: N_1 \to N$ such that $fg = f_1$.

If a foliation is a fibration, then every submanifold of a fiber is an integral manifold. Another example of an integral manifold is provided by the *slice* of a foliation. This is defined in two steps. Let (U, π) be a chart in an atlas of submersions defining the foliation $\mathscr{F}$; we write this simply as $(U, \pi) \in \mathscr{F}$. A slice of U is a connected subset L of U of the form $L = \pi^{-1}\pi(q)$, where $q \in U$. A slice of $\mathscr{F}$ is a slice of U for some $(U, f) \in \mathscr{F}$. By II,2.3 a slice is a submanifold of M, hence (L, i), i the inclusion map, is an integral manifold.

(4.2) Lemma *Let L be a slice of U. Then*:

(a) *If (N, f) is a connected integral manifold of $\mathscr{F}$, $f(N) \subset U$, and $f(N) \cap L \neq \emptyset$, then $f(N) \subset L$;*

(b) *If $(V, \sigma) \in \mathscr{F}$, then there is at most a countable number of slices of V intersecting L.*

Proof Observe first that since $D\pi f = 0$, πf is locally constant on N. Hence $f^{-1}(L \cap f(N))$ is open in N, which implies (a). It also implies that the intersection of two slices of $\mathscr{F}$ is an open subset of both. It follows from (a) that if two slices of V intersect, then they are equal. Hence if L' and L'' are two distinct slices of V intersecting L, then $L \cap L'$ and $L \cap L''$ are disjoint open subsets of L. Since L is a manifold, every family of disjoint open subsets of L is at most countable. $\square$

Our main theorem asserts that a foliated manifold is a union of leaves.

(4.3) Theorem *Let $\mathscr{F}$ be a foliation of M. Then every point of M lies in a leaf.*

Proof Let $p \in M$. We will say that a point $q \in M$ is $\mathscr{F}$-related to p if there is a finite set of slices $L_1, \ldots, L_k$ such that $p \in L_1$, $q \in L_k$, and $L_i \cap L_{i+1} \neq \emptyset$ for $i = 1, \ldots, k - 1$. Such a set of slices will be called a *chain*.

We have defined an equivalence relation among the points of M. Let $\mathscr{L}_p$ be the set of points $\mathscr{F}$-related to p. $\mathscr{L}_p$ is a union of slices and we topologize it by taking as the base all open subsets of slices contained in $\mathscr{L}_p$. (This topology differs in general from the topology of $\mathscr{L}_p$ as a subset of M. They agree if $\mathscr{L}_p$ is compact.)

Clearly, $\mathscr{L}_p$ is a connected Hausdorff space. We will show that it has a countable base.

First, notice that there is a countable atlas of submersions $\mathscr{U} = \{U_\alpha, \pi_\alpha\}$ in $\mathscr{F}$ such that every set $\pi^{-1}\pi(q)$, $q \in U_\alpha$, is connected, hence a slice. It is easy to see that if p and q are $\mathcal{F}$-related, then they are $\mathcal{F}$-related using only chains of slices from $\mathcal{U}$. Therefore if L is a slice from $\mathcal{U}$ containing p, then the last elements of all possible chains from $\mathcal{U}$ which begin with L will cover $\mathcal{L}_p$. It follows from 4.2(b) that there is at most a countable number of k-element chains from $\mathcal{U}$ with fixed first $k - 1$ elements. Hence the number of chains from $\mathcal{U}$ beginning with L is at most countable. Since each slice has a countable base, this proves that $\mathcal{L}_p$ has a countable base.

As the next step we observe that smooth structures on two intersecting slices are compatible; this also follows from II,2.3. Since $\mathcal{L}_p$ is covered by slices, it inherits a smooth structure in which slices are open subsets and smooth submanifolds. This, in turn, implies that the inclusion $i: \mathcal{L}_p \hookrightarrow M$ is a one-to-one immersion. Thus we conclude that $(\mathcal{L}_p, i)$ is an integral manifold.

To show that $\mathcal{L}_p$ is a leaf, let (N, f) be a connected integral manifold containing p. Let $\mathcal{U} = \{U_\alpha, \pi_\alpha\}$ be an atlas of submersions as before. If $q \in f(N)$, then there is a family $V_1, \ldots, V_k$ of open connected subsets of N such that $V_i \cap V_{i+1} \neq \emptyset$, $i = 1, \ldots, k - 1$, $p \in f(V_1)$, $q \in f(V_k)$, and each set $f(V_i)$ is contained in a chart U_i from $\mathcal{U}$. Since $(V_i, f \mid V_i)$ is a connected integral manifold, there is by 4.2(a) a slice L_i of U_i containing $f(V_i)$. Then $L_1, \ldots, L_k$ form a chain of slices, which means that $q \in \mathcal{L}_p$. This shows that $f(N) \subset \mathcal{L}_p$.

Both f and i are one-to-one immersions; hence there is a one-to-one map $g: N \to \mathcal{L}_p$ such that $ig = f$. Locally, the topology of $\mathcal{L}_p$ is that of a subset of M; hence g is continuous. This implies that it is an immersion. $\square$

A leaf (N, f) through a given point is unique up to a composition of f with a diffeomorphism $N \to N$. For if (N_1, f_1) and (N_2, f_2) are two intersecting leaves, then there are one-to-one immersions $h: N_1 \to N_2$, $g: N_2 \to N_1$ such that $f_2 h = f_1$ and $f_1 g = f_2$. Hence $f_2 h g = f_2$ and $f_1 g h = f_1$, which implies that h and g are both diffeomorphisms.

5 Examples

In general, it is a very hard problem to decide whether a given manifold admits a d-foliation. We give a few examples here.

(5.1) By 2.5 a nowhere vanishing vector field on M gives rise to a 1-foliation; it follows that every compact manifold with vanishing Euler-

Poincaré characteristic admits a 1-foliation with leaves diffeomorphic either to S^1 or to $\mathbf{R}^1$.

The leaves of a foliation induced by a vector field are solution curves of the corresponding differential equation; the compact leaves correspond to periodic solutions. However, it is possible that all leaves of a 1-foliation of a compact manifold are non-compact. For example, let $f: \mathbf{R}^2 \to \mathbf{R}^1, f(x, y) = x - \alpha y$, and let $\pi: \mathbf{R}^2 \to S^1 \times S^1$ be the covering map $\pi(x, y) = (x \bmod 1, y \bmod 1)$. Since Ker Df is invariant under translations of $\mathbf{R}^2$, f induces an atlas of immersions on $S^1 \times S^1$. The corresponding foliation has all leaves compact if α is rational. If α is irrational, then every leaf is non-compact and dense in $S^1 \times S^1$.

Exercise Construct a 1-foliation of $S^1 \times S^1$ with both compact and non-compact leaves present.

(5.2) There is no fibration of S^3 by 2-dimensional manifolds, but there are 2-foliations. We describe an example due to G. Reeb.

We begin by foliating $\mathring{D}^2 \times \mathbf{R}$ by surfaces $z = c + 1/(1 - x^2 - y^2)$; the corresponding 2-field is the field of tangent planes. Since this foliation is invariant under translations along $\mathbf{R}$, it induces a foliation of $\mathring{D}^2 \times S^1$ (see Fig. V,1).

Now, S^3 can be obtained by identifying two copies of $D^2 \times S^1$ along the boundaries. The foliation of S^3 is obtained by foliating the interior of each copy of $D^2 \times S^1$ in the way just described and adding one more leaf: the common boundary $S^1 \times S^1$. This becomes the only compact leaf of the resulting foliation, all other leaves are diffeomorphic to $\mathbf{R}^2$.

Another identification of boundaries of two copies of $D^2 \times S^1$ results in $S^2 \times S^1$ and a foliation of it with one compact leaf.

(5.3) Let U be an open subset of $\mathbf{R}^m$. We will show here that the well-known set of conditions for the existence of a map $f: U \to \mathbf{R}$ with a given gradient is a consequence of the Frobenius theorem, 3.2. For this purpose we will first find a sufficient condition for integrability of an $(m - 1)$-field in U.

Let $Z = (P_1, \ldots, P_m)$ be a nowhere vanishing vector field in U and E the $(m - 1)$-field of planes orthogonal to Z. By 3.2, E is completely integrable if the bracket of two vector fields X, Y in E is again in E. Let $X = \sum_i \alpha_i \partial_i$, $Y = \sum_i \beta_i \partial_i$; X and Y are in E if and only if

$$\sum_i \alpha_i P_i = 0 = \sum_i \beta_i P_i.$$

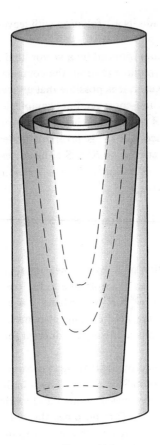

Figure V,1.

Now, $[X, Y] = \sum_{i,j} (\alpha_i \, \partial_i \beta_j - \beta_i \, \partial_i \alpha_j) \, \partial_j$; hence $[X, Y]$ is in E if and only if

$$0 = \sum_{i,j} (\alpha_i \, \partial_i \beta_j - \beta_i \, \partial_i \alpha_j) P_j = \sum_i \left(\alpha_i \sum_j P_j \, \partial_i \beta_j - \beta_i \sum_j P_j \, \partial_i \alpha_j \right)$$

$$= - \sum_i \left(\alpha_i \sum_j \beta_j \, \partial_i P_j - \beta_i \sum_j \alpha_j \, \partial_i P_j \right)$$

$$= - \sum_{i<j} (\alpha_i \beta_j - \alpha_j \beta_i)(\partial_i P_j - \partial_j P_i).$$

It follows that $\partial_i P_j = \partial_j P_i$, $i, j = 1, \ldots, m$, is a sufficient condition for E to be completely integrable.

We return now to the problem of finding $f: U \to \mathbf{R}$ with the given gradient $(P_1, \ldots, P_m)$. Let $Z = (P_1, \ldots, P_m, 1)$ be a vector field in $U \times \mathbf{R}$ and E the m-field of planes orthogonal to Z. Suppose that E is completely integrable, and let L be the leaf through $(x, t) \in U \times \mathbf{R}$ of the foliation determined by E. The projection $U \times \mathbf{R} \to U$ is a local diffeomorphism on L; hence there is a neighborhood V of x in U and a function $f: V \to \mathbf{R}$ such that L is the graph of f. This implies that the vector field N normal to L, $N = (\partial_1 f, \ldots, \partial_m f, 1)$, is parallel to Z. But if $N = \lambda Z$, then $\lambda = 1$; hence $N = Z$, i.e., $\partial_i f = P_i$, $i = 1, \ldots, m$. Thus the complete integrability of E is sufficient for the existence of f with $\nabla f = (P_1, \ldots, P_m)$. Since the P_i do not depend on the coordinate t, the sufficient condition for complete integrability is, again, $\partial_i P_j = \partial_j P_i$, $i, j = 1, \ldots, m$.

The reader might observe that while this is undoubtedly the most complicated proof of a rather elementary theorem, it can be generalized to give conditions for the existence of a map $\mathbf{R}^m \to \mathbf{R}^n$ with an *a priori* given differential.

Exercise Show that if U is simply connected, then f is defined in U. (*Hint*: The projection $U \times \mathbf{R} \to U$ restricted to L is a covering map.)

VI

Operations on Manifolds

In this chapter we describe various operations on manifolds: connected sum, attachment of handles, and surgery. All of these are special cases of a general construction, joining of two manifolds along a submanifold, presented in Sections 4 and 5. However, since all important features are already present in the special cases of connected sum and connected sum along the boundary, we discuss these two cases first in Sections 1 and 3, respectively.

The general construction is specialized to attaching of handles in Section 6. We are particularly interested in the question when the attachment of two handles of consecutive dimensions results in no change to the manifold, that is when the second handle "destroys" the first. The main result in this direction, Smale's Cancellation Lemma, is proved in Section 7. The proof is based on an elementary but far-reaching theorem concerning attachment of disc bundles along a cross section in the boundary.

In Section 8 we look at handle attachment from a different point of view, more convenient for homology calculations. Section 9 introduces the operation of surgery, and in Section 10 we calculate some related homological results. In Section 11 we define handlebodies and investigate their structure. Some important examples are constructed in Section 12 using the plumbing construction. The results of the last two sections will not be used until Chapter VIII.

1   Connected Sum

Connected sum is the operation of "joining two manifolds by a tube."

Given two connected m-dimensional manifolds M_1, M_2, let $h_i: \mathbf{R}^m \to M_i$, $i = 1, 2$, be two imbeddings. If both manifolds are oriented, then we assume that h_1 preserves the orientation and h_2 reverses it.

Let $\alpha: (0, \infty) \to (0, \infty)$ be an arbitrary orientation reversing diffeomorphism. We define $\alpha_m: \mathbf{R}^m - \mathbf{0} \to \mathbf{R}^m - \mathbf{0}$ by

$$\alpha_m(v) = \alpha(|v|) \frac{v}{|v|}.$$

The *connected sum* $M_1 \# M_2(h_1, h_2, \alpha)$ is the space obtained from the (disjoint) union of $M_1 - h_1(\mathbf{0})$ and $M_2 - h_2(\mathbf{0})$ by identifying $h_1(v)$ with $h_2(\alpha_m(v))$ (see Fig. VI,1).

Recall that if A, B are two spaces and f maps a subset of A to B, then $A \cup_f B$ stand as for the identification space obtained from the disjoint union $A \cup B$ by identifying x with $f(x)$ ([Du,VI]). With this notation

$$M_1 \# M_2(h_1, h_2, \alpha) = (M_1 - h_1(\mathbf{0})) \cup_g (M_2 - h_2(\mathbf{0})), \qquad g = h_2 \alpha_m h_1^{-1}.$$

In general we will not specify h_1, h_2, α. This is justified by the following:

(1.1)   Theorem $M_1 \# M_2$ *is a smooth manifold, connected if $m > 1$ and oriented if both M_1, M_2 are oriented. It does not depend—up to diffeomorphism—on the choice of α and of the imbeddings h_i.*

Proof   It follows immediately from the Invariance of Domain that:

(∗)   The projections $M_i - h_i(\mathbf{0}) \to M_1 \# M_2$, $i = 1, 2$, are open maps.

This implies that $M_1 \# M_2$ is second countable. We have to show that it is a Hausdorff space. (This is not immediate: The identification of $h_1(x, t)$ with $h_2(x, t)$ yields in general a non-Hausdorff manifold!) In view of (∗) this reduces to showing that if $x \in M_1 - h_1(\mathbf{0})$ and $y \in M_2 - h_2(\mathbf{0})$ have distinct images in $M_1 \# M_2$, then they have neighborhoods with disjoint images. The verification is a routine case-by-case checking.

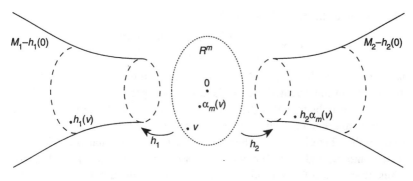

$M_1 - h_1(0)$ R^m $M_2 - h_2(0)$

0
$\cdot$
$\cdot \alpha_m(v)$

$\cdot h_1(v)$ $\cdot h_2\alpha_m(v)$

$\cdot v$

h_1 h_2

Figure VI,1

Now, the fact that $g_2\alpha_m h_1^{-1}$ is a (orientation preserving) diffeomorphism together with (*) implies that the smooth structures on $M_1 - h_1(0)$ and $M_2 - h_2(0)$ are compatible, hence yield a (oriented) smooth structure on $M_1 \# M_2$. (This is the unique structure for which projections in (*) are diffeomorphisms.)

We show now that $M_1 \# M_2$ does not depend on the choice of the imbeddings h_1, h_2.

Choose t_0, t_1 so that $0 < t_0 < t_1 < 1$, and let $\mathbf{R}^m(t_0, t_1) = \{v \in \mathbf{R}^m \,|\, t_0 < |v| < t_1\}$. Note that $\alpha_m(\mathbf{R}^m(t_0, t_1)) = \mathbf{R}^m(\alpha(t_1), \alpha(t_0))$. Now, remove from $h_1(\mathbf{R}^m)$ the closed disc of radius t_0 and from $h_2(\mathbf{R}^m)$ the closed disc of radius $\alpha(t_1)$, and glue $h_1(\mathbf{R}^m(t_0, t_1))$ to $h_2(\mathbf{R}^m(\alpha(t_1), \alpha(t_0))$ via the diffeomorphism $h_2\alpha_m h_1^{-1}$. This yields a manifold $M = M(h_1, h_2, \alpha)$. Clearly:

(**) $M(h_1, h_2, \alpha)$ is diffeomorphic to $M_1 \# M_2(h_1, h_2, \alpha)$.

Notice now that M depends only on $h_2\alpha_m h_1^{-1}$ restricted to $h_1(\mathbf{R}^m(t_0, t_1))$. Thus if β_1 shrinks $\mathbf{R}^m$ to an open disc $D^m(u)$ with $u > t_1$ and is the identity in $\mathbf{R}^m(t_1)$, and β_2 is defined similarly, then $M(h_1\beta_1, h_2\beta_2, \alpha) = M(h_1, h_2, \alpha) = M_1 \# M_2$. But $h_i\beta_i$ imbeds $\mathbf{R}^m$ as a proper tubular neighborhood of $h_i\beta_i(0)$. This means that we can always assume that $h_i(\mathbf{R}^m)$ is a proper tubular neighborhood of $h_i(0)$, $i = 1, 2$.

This, in turn, easily implies that:

(***) $M_1 \# M_2(h_1, h_2, \alpha)$ does not depend on the choice of h_1, h_2.

For suppose that h' and h_1 both imbed $\mathbf{R}^m$ as a proper tubular neighborhood of 0. Then III,3.5 yields a diffeomorphism $g: M_1 \to M_1$ such that

$gh_1 = h'$, and the map G defined by

$$G(x) = \begin{cases} g(x) & \text{if } x \in M_1 - h_1(\mathbf{0}), \\ x & \text{if } x \in M_2 - h_2(\mathbf{0}), \end{cases}$$

is a diffeomorphism of $M_1 \# M_2(h', h_2, \alpha)$ onto $M_1 \# M_2(h_1, h_2, \alpha)$. Moreover, in the oriented case G is orientation preserving.

It remains to be seen that the choice of α is immaterial. For this, suppose that β is another orientation reversing diffeomorphism of the ray $(0, \infty)$ onto itself. There is then a diffeomorphism g of $(0, \infty)$ that is the identity near 0 and ∞, and such that $\alpha = \beta g$ in some segment (t_0, t_1) (cf. III,3.6). Thus, there is a diffeomorphism g_m of $\mathbf{R}^m$ onto itself such that $\alpha_m = \beta_m g_m$ in $\mathbf{R}^m(t_0, t_1)$. Now,

$$M_1 \# M_2(h_1, h_2, \alpha) = M_1 \# M_2(h_1 g_m, h_2, \alpha) \qquad \text{(by } (***))$$

$$= M(h_1 g_m, h_2, \alpha) \qquad \text{(by } (**))$$

$$= M(h_1, h_2, \beta)$$

since both manifolds are obtained by the same identification: In $h_1(\mathbf{R}^m(t_0, t_1))$ we have $h_2 \alpha_m (h_1 g_m)^{-1} = h_2 \beta_m h_1^{-1}$. $\square$

The strength of 1.1 is in allowing us to use arbitrary—not necessarily proper—imbeddings to construct connected sums. This is exploited in the following situation: Let $h_1: \mathbf{R}^m \to \mathbf{R}^m$ be the identity map and let $h_2: \mathbf{R}^m \to S^m$ be an imbedding equivariant rel. the action of $\mathbf{O}(m + 1)$ on S^m which keeps $h_2(\mathbf{0})$ fixed, thus effectively the action of $\mathbf{O}(m)$. Then $\mathbf{O}(m)$ acts on $\mathbf{R}^m \# S^m$.

(1.2) Proposition   *There is a diffeomorphism $\mathbf{R}^m \# S^m \to \mathbf{R}^m$ that is equivariant with respect to this action, and an identity outside of a compact set.*

Proof   Let $a_\pm = (0, \dots, 0, \pm 1) \in S^m \subset \mathbf{R}^{m+1}$ and let $p_\pm: \mathbf{R}^m \to S^m$ be the projection from $a_\pm$ (i.e., the inverse of the projection $h_\pm$ from I,1.2). Then, $p_\pm$ is a diffeomorphism onto $S_m - \{a_\pm\}$ and p_- reverses the orientation. Let $h_1: \mathbf{R}^m \to \mathbf{R}^m$ be the identity map, let $h_2 = p_-$, and $\alpha(t) = 1/t$.

Now, the diffeomorphism h of $\mathbf{R}^m \# S^m(h_1, h_2, \alpha)$ onto $\mathbf{R}^m$ is defined to be the identity map in $\mathbf{R}^m - h_1(\mathbf{0})$ and p_+^{-1} in $S^m - \{a_+\}$. This works because $p_-^{-1} p_+ = \alpha_m$. $\square$

The geometric idea of the connected sum as two manifolds joined by a tube is visible in the following construction.

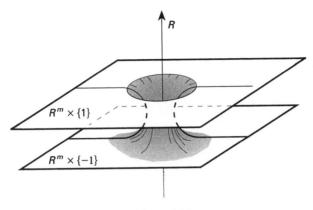

Figure VI,2

Let $\mathbf{R}_0^m$, $\mathbf{R}_1^m$ be two copies of $\mathbf{R}^m$ but with opposite orientations and let $h_i: \mathbf{R}^m \to \mathbf{R}_i^m$, $i = 0, 1$, be imbeddings of $\mathbf{R}^m$ as interior of the unit disc.

(1.3) Proposition *There is an imbedding h of $\mathbf{R}^m \# \mathbf{R}^m(h_1, h_2, \alpha)$ in $\mathbf{R}^m \times [-1, 1]$ that imbeds $h_1(\mathbf{R}^m - 0) \cup h_2(\mathbf{R}^m - 0)$ as "a tube" in $\mathbf{R}^m \times [-1, 1]$ and such that $h(x) = (x, (-1)^i)$ elsewhere in $\mathbf{R}_i^m$ (see Fig. VI,2).*

Proof If $m = 1$, then one starts with the imbedding of $\mathbf{R}_0^1 \# \mathbf{R}_1^1$ as the hyperbola $3x^2 - y^2 = 1$ and then brings it into the desired shape by sending each point (x, y) on the hyperbola to $(x, y/g(x))$, where g is a smooth positive function equal to $(3x^2 - 1)^{1/2}$ for $x^2 \geq 1$ and $\leq \sqrt{2}$ elsewhere. Rotation of this imbedding around the y axis produces the desired imbedding for $m > 1$. $\square$

Observe now that if $\mathbf{R}^m \# (-\mathbf{R}^m)$ is so imbedded in $\mathbf{R}^m \times [-1, 1]$, then it bounds a manifold that has $\mathbf{R}^m$ with the interior of a disc deleted as a deformation retract. The deformation simply moves points on "vertical" lines. This has the following consequence, which we will use in Chapter VIII. Let M be an oriented manifold, $h: \mathbf{R}^m \to M$ an imbedding. The connected sum $M \# (-M)$ can now be imbedded in $M \times [-1, 1]$ by imbedding $h(\mathbf{R}^m) \# h(\mathbf{R}^m)$ in $h(\mathbf{R}^m) \times [-1, 1]$ as in 1.3, and the rest in the obvious way. The resulting manifold will bound a manifold that has M with the interior of a disc deleted as a deformation retract.

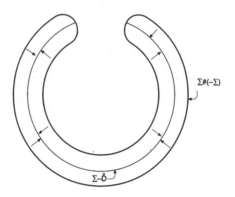

$\Sigma\#(-\Sigma)$

$\Sigma - \overset{\circ}{\mathrm{b}}$

Figure VI,3

In particular, if Σ is a homotopy sphere, then Σ with the interior of a disc deleted is a contractible manifold. Therefore:

(1.4) Corollary *If Σ is a homotopy sphere, then $\Sigma \# (-\Sigma)$ bounds a contractible manifold (see Fig. VI,3).*

Exercise If M and N are disjoint submanifolds of codimension >1 of a connected manifold W, then $M \# N$ imbeds in W.

2 # and Homotopy Spheres

To calculate the homology of $M_1 \# M_2$ one applies the Mayer–Vietoris sequence to the pair (A_1, A_2), where A_i is the image in $M_1 \# M_2$ of $M_i - h_i(\mathbf{0})$, $i = 1, 2$. Since $A_1 \cap A_2$ has the homotopy type of S^{m-1} we obtain immediately that $H_i(M_1 \# M_2) \simeq H_i(M_1) \oplus H_i(M_2)$ for $0 < i < m$, at least when both M_1, M_2 are closed and oriented.

Analogously, the Seifert–Van Kampen theorem applied to the same pair shows that, for $m \geq 3$, $\pi_1(M_1 \# M_2) \simeq \pi_1(M_1) \times \pi_1(M_2)$. Taken together, this yields:

(2.1) Proposition *The connected sum of two manifolds is a homotopy sphere if and only if both are homotopy spheres.*   □

(Actually, an additional argument is needed in the case $m = 2$.)

Exercise Calculate $\pi_1(M_1 \# M_2)$ when M_1, M_2 are 2-dimensional closed manifolds.

(2.2) Theorem *The set of connected, oriented, and closed m-dimensional manifolds is, under the operation of connected sum, an associative and commutative monoid with identity.*

Proof We have to show that:

(a) $M_1 \# M_2 = M_2 \# M_1$;
(b) $(M_1 \# M_2) \# M_3 = M_1 \# (M_2 \# M_3)$;
(c) $M \# S^m = M$.

To verify (a), let $h: \mathbf{R}^m \to \mathbf{R}^m$ be the reflection in the first coordinate and suppose that $M_1 \# M_2$ is constructed using imbeddings h_1, h_2. To construct $M_2 \# M_1$ we can use the imbeddings $h_i h$, $i = 1, 2$. A diffeomorphism $M_1 \# M_2 \to M_2 \# M_1$ is then constructed by requiring it to be the identity on $M_1 - h_1(0)$ and $M_2 - h_2(0)$.

The proof of (b) is left to the reader.

Finally, (c) is an immediate consequence of 1.2. $\square$

Of course, 2.2 holds without assuming the manifolds are oriented.

The structure of these monoids is not known beyond $m = 2$. For $m = 2$ it is known that, up to a diffeomorphism, a given topological 2-manifold can carry only one smooth structure. The topological classification of 2-manifolds implies then the following:

(2.3) Theorem *The monoid of 2-dimensional closed, compact connected manifolds is generated by the torus T and the projective plane P with the relation $3P = T \# P$. The monoid of oriented 2-manifolds is isomorphic to the monoid of natural numbers.*

Consider now the monoid of oriented compact connected closed m-dimensional manifolds. The subset A^m of invertible elements is a group. It follows from 2.1 that elements of A^m are homotopy spheres. In fact, a stronger assertion is true: Elements of A^m are topological spheres. This is a consequence of the following:

(2.4) Proposition *If $M \# N$ is homeomorphic to S^m, then M is homeomorphic to S^m.*

Proof Let $D \subset M$ be an imbedded m-disc, S its boundary and D' the closure of its complement. We regard $M \# N$ as constructed using an imbedding of $\mathbf{R}^m$ in the interior of D. Let $h: M \# N \to S^m$ be a diffeomorphism. By a theorem of B. Mazur ([Ma1], [Bd, IV, 19.11]) the closure of each component of $h(S)$ is homeomorphic to an m-disc. Therefore D' is homeomorphic to an m-disc, and M is a union of two (topological) m-discs with identified boundaries, hence is homeomorphic to S^m.   □

Another proof, due to J. Stallings, is based on the possibility of defining an infinite connected sum $M_1 \# M_2 \# M_3 \# \cdots$ that is associative and satisfies

$$S^m \# S^m \# S^m \# \cdots = \mathbf{R}^m.$$

Assuming this verified, we have

$$\mathbf{R}^m = S^m \# S^m \# S^m \# \cdots = (M \# N) \# (M \# N) \# (M \# N) \# \cdots$$
$$= M \# (N \# M) \# (N \# M) \cdots$$
$$= M \# S^m \# S^m \# S^m \# \cdots = M \# \mathbf{R}^m.$$

All that remains now is a simple exercise:

Exercise   Show that if $M \# \mathbf{R}^m$ is diffeomorphic to $\mathbf{R}^m$, then M is homeomorphic to S^m.

Note that this proof yields a weaker theorem: we have to assume that $M \# N$ is diffeomorphic to S^m.

By 2.4 the group A^m can be construed as the group of invertible differential structures on the topological m-dimensional sphere. In VIII,5 we shall show that for $m \geq 5$ all differential structures on spheres are invertible. (This is also true for $m < 5$ but our methods do not apply.) In fact, we shall prove a much stronger result: For $m \geq 5$, A^m coincides with the submonoid of all homotopy spheres. Whether this last statement is true for $m = 3$ is not known.

Exercise   Let D be a closed m-disc in M^m and β the involution on ∂D interchanging the antipodal points. The space P is obtained from $M - \text{Int } D$ by identifying every pair $x, \beta(x)$ to a point. Show that $P = M \# P^m$.

3 Boundary Connected Sum

Next in order we consider the operation of boundary connected sum. In this case two manifolds with boundary are joined along a disc in the boundary. This is done as follows.

Let M_1, M_2 be two connected m-dimensional manifolds with connected boundaries, let $h_i \colon \mathbf{R}^{m-1} \to \partial M_i$, $i = 1, 2$, be two imbeddings and let $\bar{h}_i \colon \mathbf{R}^m_+ \to M_i$ be imbeddings extending the h_i where $\mathbf{R}^m_+ = \{x \in \mathbf{R}^m \,|\, x_m \geq 0\}$. As before, we assume that if M_1, M_2 are both oriented, then h_1 preserves orientation and h_2 reverses it.

The *boundary connected sum* of M_1 and M_2, denoted $M_1 \#_b M_2$, is the space obtained from the (disjoint) union of $M_1 - h_1(0)$ and $M_2 - h_2(0)$ by identifying $\bar{h}_1(v)$ with $\bar{h}_2(\alpha_m(v))$.

(3.1) Theorem $M_1 \#_b M_2$ *is a smooth connected manifold, oriented if both M_1, M_2 are oriented. It does not depend—up to diffeomorphism—on the choice of α and of the imbeddings h_i. Moreover,*

$$\partial(M_1 \#_b M_2) = \partial M_1 \# \partial M_2.$$

Proof The proof is a word-for-word repetition of the proof of 1.1. The last part follows from the fact that the construction of $M_1 \#_b M_2$ restricted to the boundary is precisely that used to construct the connected sum. $\square$

Theorem 2.2 can be restated in the context of boundary connected sum. Corresponding to 2.2(c), we have

(3.2) $$M \#_b D^m = M.$$

This follows immediately from the following analogue of 1.2:

(3.3) Proposition *There is a diffeomorphism $\mathbf{R}^{m+1}_+ \#_b D^{m+1} \to \mathbf{R}^{m+1}_+$ that is the identity outside of a compact set.*

Proof As h_1 and $\bar{h}_1$ we take the identity maps. As h_2 we take the projection $\mathbf{R}^m \to S^m - a_-$ from a_-, i.e., for $v \in \mathbf{R}^m$,

$$h_2(v) = \frac{(2v, 1 - v^2)}{v^2 + 1} \in \mathbf{R}^m \times \mathbf{R} = \mathbf{R}^{m+1}.$$

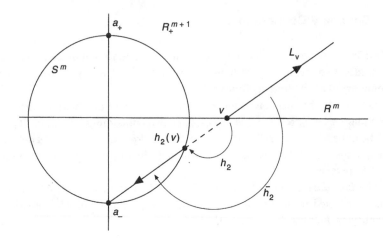

Figure VI,4

The extension $\bar{h}_2$ to $\mathbf{R}^m \times \mathbf{R}_+$ is now defined by

$$\bar{h}_2(v, t) = \frac{(2v, 1 - v^2 - t^2)}{v^2 + (1 + t)^2}, \qquad (v, t) \in \mathbf{R}^m \times \mathbf{R}_+.$$

(If L_v is the ray beginning at a_- through $v \in \mathbf{R}^m$, then $\bar{h}_2$ sends $L_v \cap \mathbf{R}_+^{m+1}$, which is a ray beginning at v, onto the segment $[h_2(v), a_-]$. See Fig. VI,4.)

Now, $\mathbf{R}_+^{m+1} \#_b D^{m+1}$ is obtained by identifying $v \in \mathbf{R}_+^{m+1} - 0$ with $\bar{h}_2(\alpha_{m+1}(v))$, where $\alpha(t) = 1/t$.

We define the diffeomorphism $h: \mathbf{R}_+^{m+1} \#_b D^{m+1} \to \mathbf{R}_+^{m+1}$ to be the identity map on $\mathbf{R}_+^{m+1} - \{0\}$ and the map

$$(v, t) \mapsto \frac{(2v, 1 - v^2 - t^2)}{v^2 + (1 - t)^2}$$

on $D^{m+1} - \{a_+\}$.   □

The geometric argument in this proof is "one-half" of that in 1.2: Let $S_+ = S^m \cap \mathbf{R}_+^{m+1}$ and consider that part of $\mathbf{R}^m \# S^m$ where $x_m \geq 0$. This is easily seen to be $\mathbf{R}_+^m \#_b S_+$. (The case $m = 2$ is shown in Fig. VI,5.) Since the diffeomorphism h maps it onto $\mathbf{R}_+^m$, and S_+ is diffeomorphic to D^m, 3.3 follows.

Exercise   Show that $M_1 \#_b M_2$ has the homotopy type of $M_1 \vee M_2$.

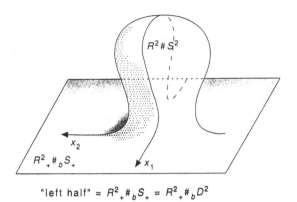

"left half" = $R^2_+ \#_b S_+ = R^2_+ \#_b D^2$

Figure VI,5

4 Joining Manifolds along Submanifolds

Both operations—connected sum and boundary connected sum—consist in joining two manifolds by means of a diffeomorphism of tubular neighborhoods of points. Viewing a point as a submanifold it is natural to generalize this to an operation of joining two manifolds along tubular neighborhoods of submanifolds. We again distinguish two cases according to whether we join manifolds along submanifolds of the interior or along submanifolds of the boundary.

Suppose we are given two $(n + k)$-dimensional manifolds M_1, M_2 and a k-dimensional Riemannian vector bundle over an n-dimensional closed, compact manifold N with total space E. Its zero section will be identified with N. If $\alpha : (0, \infty) \to (0, \infty)$ is an orientation reversing diffeomorphism, we define $\alpha_E : E - N \to E - N$ by

$$\alpha_E(v) = \alpha(|v|)\frac{v}{|v|}.$$

Now, if h_1, h_2 are two imbeddings of E in the interiors of M_1, M_2, respectively, then a new manifold $M(h_1, h_2)$ is obtained by identifying $v \in h_1(E - N)$ with $h_2\alpha_E h_1^{-1}(v)$. That it is a manifold and that the operation does not depend on α is verified exactly as in 1.1. Observe that $h_i(E)$ is necessarily a tubular neighborhood of $h_1(N)$, $i = 1, 2$.

We will refer to this operation as *pasting* of two manifolds along submanifolds. It is a generalization of connected sum: It is the connected sum when

N is a single point (except that we do not specify orientations). If N_1 and N_2 are given diffeomorphic submanifolds of M_1 and M_2 with isomorphic tubular neighborhoods, then it is possible to paste M_1 and M_2 along them.

An important special case is that in which N is the standard n-sphere S^n in S^m, that is, we paste a manifold M and S^m along an imbedded n-sphere $S \subset M$ and $S^n \subset S^m$. (S must have a trivial normal bundle.) The operation is then called a *surgery*, or a *spherical modification*, on the sphere S. We will return to this in Section 8.

There is no difficulty in extending the operation of pasting two manifolds along a submanifold to the case where N is a manifold with boundary: We require that $h_i | N$ be neat imbeddings and that $h_i(E)$ be neat tubular neighborhoods of $h_i(N)$, $i = 1, 2$. In particular, if L_1 and L_2 are neatly imbedded arcs in M_1, M_2 respectively, then their normal bundles are certainly trivial and it is possible to paste M_1 and M_2 along L_1 and L_2.

Exercise Let M_1, M_2 be two k-sphere bundles over closed manifolds N_1, N_2 respectively. Choose fibers S_1 in M_1, S_2 in M_2. Then the manifold resulting from pasting M_1 and M_2 along S_1, S_2 is a k-sphere bundle over $N_1 \# N_2$.

5   Joining Manifolds along Submanifolds of the Boundary

The last operation we shall consider here is the operation of joining two manifolds along imbedded submanifolds in the boundary. This will generalize the boundary connected sum.

Suppose we are given two $(n + k + 1)$-dimensional manifolds M_1, M_2 and a k-dimensional vector bundle ξ over an n-dimensional closed compact manifold N. Suppose that h_1, h_2 are two imbeddings of its total space E in ∂M_1, ∂M_2, respectively, and that $\bar{h}_1$, $\bar{h}_2$ are extensions of h_1, h_2 as tubular neighborhoods of $h_1(N)$, $h_2(N)$ in M_1, M_2. That is, $\bar{h}_1$ and $\bar{h}_2$ imbed "one-half" of the $(k + 1)$-dimensional bundle $\xi \oplus \varepsilon^1$, *cf.* III,4. Now, if we let E' be its total space, then a new manifold is obtained by identifying $v \in \bar{h}_1(E' - N)$ with $\bar{h}_2 \alpha_{E'} \bar{h}_1^{-1}(v)$.

We will refer to this manifold as M_1, M_2 joined along submanifolds in the boundary and denote it $M(h_1, h_2)$.

The operation depends on the choice of imbeddings h_1, h_2, but not on the choice of extensions $\bar{h}_1$, $\bar{h}_2$; this is reflected in the notation. (As before, it is enough to show this for proper tubular neighborhoods. In this case it is a version of the Uniqueness of Collars Theorem and it follows from III,3.1 by an argument analogous to that used to prove III,3.3.) The dependence on h_1, h_2 is given by the following:

(5.1) Proposition *Let* $M = M(h_1, h_2), M' = M(h_1', h_2')$, *where* $h_i | N = h_i' | N, i = 1, 2$. *Then there is an automorphism* g *of* E *such that* $M' = M(h_1 g, h_2)$.

In other words, what can be achieved by modifying both imbeddings can also be achieved by modifying only one, by composing it with an automorphism of E. This follows from III,3.1 and the fact that α_E commutes with automorphisms of E. Details are left as an exercise.

A special case of this operation is when $N = \partial M_1$, h_1 is the identity map and $h_2: \partial M_1 \to \partial M_2$ a diffeomorphism. Given collars $\partial M_1 \times \mathbf{R}_+ \subset M_1, \partial M_2 \times \mathbf{R}_+ \subset M_2$, we obtain a new manifold by identifying (x, t) with $(x, 1/t)$. However, it is more convenient to view this new manifold as $M_1 \cup_{h_2} M_2$, because this manifold contains M_1, M_2 as subsets. There is an obvious homeomorphism between the two that we use to give $M_1 \cup_{h_2} M_2$ a smooth structure.

An example of this is the double of M: in this case M_1 and M_2 are two copies of M and h is the identity map on their boundaries.

In the case of oriented manifolds appropriate conditions on the orientability have to be added. For instance, to obtain the double as an oriented manifold we take M_2 to be M_1 with the opposite orientation.

We now consider an important special case. Let $h: \partial D^m \to \partial D^m$ be an orientation preserving diffeomorphism and let $\Sigma(h) = D^m \cup_h (-D^m)$. We leave as an exercise the proof of the following lemma.

(5.2) Lemma $\Sigma(h)$ *is diffeomorphic to* S^m *if and only if* h *extends over* D^m. *Moreover,* $\Sigma(hg) = \Sigma(h) \# \Sigma(g)$. □

In other words, there is a monomorphism from the group Γ^m to A^m (*cf.* III,6.2). In VIII,5 we will show that for $m \geq 5$ it is surjective.

Exercise Let $g: D^m \to M$ be an imbedding, $h: \partial D^m \to \partial D^m$ a diffeomorphism, and let the manifold M' be obtained from M by removing the interior of $g(D^m)$ and then gluing it back using gh, i.e., $M' = (M - \mathrm{Int}(g(D^m))) \cup_{gh} D^m$. Show that M' is diffeomorphic to $M \# \Sigma(h)$.

The operation of joining two manifolds along submanifolds in the boundary generalizes the connected sum along the boundary. There is a corresponding generalization of 3.2, in which the disc D^m is replaced by a disc bundle, and it is of importance.

We consider $M(h_1, h_2)$ obtained by joining M_1 with M_2, where M_2 is a closed disc bundle over a manifold N. Moreover we assume that they are joined along a section in the boundary, that is, that $h_2 | N$ maps N to ∂M_2 as a cross section of the bundle.

(5.3) Proposition   *Under these assumptions $M(h_1, h_2)$ is diffeomorphic to M_1.*

Proof   The proof is a straightforward generalization of the proof of 3.3. Let $s: N \to M_2$ be the section in question, η the 1-dimensional subbundle generated by s and $\eta^\perp$ its orthogonal complement, all as in III,4. We take as the tubular neighborhood h_2 of $s(N)$ in ∂M_2 the map p in III,4.3, i.e., the map

$$v \mapsto \frac{2}{1+v^2} v + \frac{1-v^2}{1+v^2} s(x)$$

where v is in the fiber of $\eta^\perp$ over x.

The extension of h_2 to a tubular neighborhood of s in M_2 is defined, as in III,4.4, by

$$\bar{h}_2(v, t) = \frac{2}{v^2 + (1-t)^2} v + \frac{1-v^2-t^2}{v^2+(1+t)^2} s(x), \qquad (v, t) \in \mathbf{R}^m \times \mathbf{R}_+.$$

Now, with $\bar{h}_1$ an arbitrary tubular neighborhood in M_1 of $h_1 | N$, the diffeomorphism $h: M(h_1, h_2) \to M_1$ is defined to be the identity on $M_1 - h_1(N)$ and $\bar{h}_1 g$ on $M_2 - s(N)$, where

$$g(v + ts(x)) \mapsto \frac{2}{v^2 + (1-t)^2} v + \frac{1-v^2-t^2}{v^2+(1-t)^2} s(x).$$

This proves 5.3 with arbitrary h_1, and h_2 chosen as in the preceding. But by 5.1 this is the general case. $\square$

As we have already said, 3.2 is a special case of this proposition. Another special case is the diffeomorphism $M \cup_h (\partial M \times I) \to M$, where h is arbitrary.

6 Attaching Handles

A particularly important case of the operation of joining two manifolds along a submanifold is that in which one of them is a disc and they are joined along a sphere. For this case, we will establish a special notation.

Let $m = \lambda + \mu$. If $x \in \mathbf{R}^m = \mathbf{R}^\lambda \times \mathbf{R}^\mu$, we write $x = (x_\lambda, x_\mu)$, i.e., $x_\lambda \in \mathbf{R}^\lambda, x_\mu \in \mathbf{R}^\mu$ stand for projections of x. With this notation $S^{\mu-1} = \{x \in D^m \,|\, x_\lambda = \mathbf{0}, x_\mu^2 = 1\}$. Let $1 > \varepsilon \geq 0$, $T(\varepsilon) = \{x \in D^m \,|\, x_\lambda^2 > \varepsilon\}$, and let $\alpha: T(\varepsilon) - S^{\lambda-1} \to T(\varepsilon) - S^{\lambda-1}$ be given by

$$(6.1) \qquad \alpha(x_\lambda, x_\mu) = \left(\frac{x_\lambda}{|x_\lambda|} (1 - x_\lambda^2 + \varepsilon)^{1/2}, x_\mu \frac{(x_\lambda^2 - \varepsilon)^{1/2}}{(1 - x_\lambda^2)^{1/2}} \right).$$

We view $T(\varepsilon)$ as a tubular neighborhood of $S^{\lambda-1}$ in D^m with the projection $(x_\lambda, x_\mu) \mapsto x_\lambda/|x_\lambda|$; through most of this chapter we have $\varepsilon = 0$ and abbreviate $T(0) = T$.

Note that α is the composition of the diffeomorphism $D^m - S^{\lambda-1} \to \mathring{D}^\lambda \times D^\mu$ given by

$$(6.1.1) \qquad (x_\lambda, x_\mu) \mapsto (x_\lambda, x_\mu/(1 - x_\lambda^2)^{1/2})$$

with the involution on $(\mathring{D}^\lambda - \mathbf{0}) \times D^\mu$:

$$(6.1.2) \qquad (x_\lambda, x_\mu) \mapsto \left(\frac{x_\lambda}{|x_\lambda|} (1 - x_\lambda^2 + \varepsilon)^{1/2}, x_\mu \right),$$

followed by the inverse of 6.1.1.

It follows easily that α preserves those great λ-spheres in ∂D^m which contain $S^{\lambda-1}$.

We insert here a technical lemma.

Let $x \in S^\mu$. The great λ-sphere in S^{m-1} that contains $S^{\lambda-1}$ and x is divided into two hemispheres by $S^{\lambda-1}$. Let $K(x)$ stand for the hemisphere that contains x and let $K \subset K(x)$ be a λ-disc centered on x; if $x = (\mathbf{0}, x_\mu^0)$, then for some t_0,

$$K = \{(x_\lambda, x_\mu) \in S^{m-1} \,|\, x_\mu = t x_\mu^0, t_0 \leq t \leq 1\}.$$

Applying 6.1 we see that

$$\alpha(K - \{x\}) \cup S^{\lambda-1}$$
$$= \{(x_\lambda, x_\mu) \in S^{m-1} \,|\, x_\mu = (1 - t^2)^{1/2} x_\mu^0, t_0 \leq t \leq 1)\}.$$

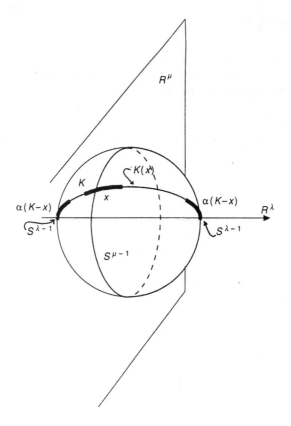

Figure VI,6

This implies (see Fig. VI,6):

(6.2) Lemma $\alpha(K - \{x\} \cup S^{\lambda-1}) = K(x) \cap T(t_0^2)$ *is a smooth manifold, a subbundle of* $T(t_0^2)$, *and a collar of* $S^{\lambda-1}$ *in* $K(x)$. *Its boundary consists of* $S^{\lambda-1}$ *and* $\alpha(\partial K)$.   □

We now define the operation of attaching handles.

Let $h: S^{\lambda-1} \to \partial M^m$ be an imbedding, and let $\bar{h}: T \to M^m$ be an extension of h and a tubular neighborhood of $h(S^{\lambda-1})$ in M^m. Then the manifold M_1 obtained from $M^m - h(S^{\lambda-1})$ and $D^m - S^{\lambda-1}$ by identifying $x \in T - S^{\lambda-1}$ with $\bar{h}\alpha(x)$ will be referred to as M *with the handle attached along* $h(S^{\lambda-1})$

and denoted, symbolically, $M_1 = M \cup H^\lambda$; $h(S^{\lambda-1})$ will be called the *attaching sphere*.

We will identify $M^m - h(S^{\lambda-1})$ and $D^m - S^{\lambda-1}$ with their images in $M \cup H^\lambda$. In particular, $D^m - S^{\lambda-1}$ (as a subset of M_1) will be called the *handle*, the μ-disc $D^\mu = D^m \cap \mathbf{R}^\mu$ the *belt disc*, and its boundary $S^{\mu-1}$ the *belt sphere*.

Observe that if M is oriented and $\bar{h}$ reverses orientation, then $M \cup H^\lambda$ admits an orientation agreeing with the orientation of M and with the orientation of the handle derived from the standard orientation of D^m.

Attaching a 0-handle to M means taking a disjoint union of M with D^m. The following exercise describes attaching a 1-handle.

Exercise Let M_1, M_2 be two connected manifolds with boundary and let $x_i \in \partial M_i$, $i = 1, 2$. Let M be $M_1 \cup M_2$ with a 1-handle attached along the 0-sphere $\{x_1, x_2\}$. Then $M = M_1 \#_b M_2$.

We now study the effects of attaching a handle to a disc.

(6.3) Proposition *If M is obtained by attaching a λ-handle to D^m along $S^{\lambda-1}$, then M is a μ-disc bundle B over a manifold Σ homeomorphic to the sphere S^λ.*

Proof Let D_1, D_2 be two copies of the disc D^m. The subscript will identify to which disc we refer: e.g., T_1 will be the tubular neighborhood of $S_1^{\lambda-1}$ in D_1.

The construction of M begins with a given diffeomorphism h of a tubular neighborhood $T_1(\varepsilon) \cap \partial D_1$ of $S_1^{\lambda-1}$ onto a tubular neighborhood of $S_2^{\lambda-1}$ in ∂D_2. This is extended to a diffeomorphism $\bar{h}$ of $T_1(\varepsilon)$ onto a tubular neighborhood of $S_2^{\lambda-1}$ in D_2; then x is identified with $\bar{h}\alpha(x)$. To obtain a representation of M as a disc bundle, we note that by III,3.5 we can assume that for some $\gamma \varepsilon \pi_{\lambda-1}(\mathbf{0}(\mu))$

(6.3.1) $$h(x_\lambda, x_\mu) = \left(|x_\lambda| h\!\left(\frac{x_\lambda}{|x_\lambda|}\right), \gamma\!\left(\frac{x_\lambda}{|x_\lambda|}\right) \cdot x_\mu\right).$$

We can now take $\varepsilon = 1$ and as $\bar{h}$ the radial extension

$$(x_\lambda, x_\mu) \mapsto \left(|x_\lambda| h\!\left(\frac{x_\lambda}{|x_\lambda|}\right), x_\mu\right)$$

of h followed by an automorphism given by γ. Thus if we identify $D_i^m - S_i^{\lambda-1}$

with $\mathring{D}_i^\lambda \times D^\mu$, $i = 1, 2$, via the diffeomorphism 6.1.1, then M is diffeomorphic to $B = (\mathring{D}_1^\lambda \times D^\mu) \cup_\sim (\mathring{D}_2^\lambda \times D^\mu)$ with the identification

$$(6.3.2) \quad (x_\lambda, x_\mu) \sim \left(h\left(\frac{x_\lambda}{|x_\lambda|}\right)(1 - x_\lambda^2)^{1/2}, \gamma\left(\frac{x_\lambda}{|x_\lambda|}\right) \cdot x_\mu \right), \qquad \gamma \in \pi_{\lambda-1}(\mathbf{O}(\mu)).$$

This identification commutes with the projection $(x_\lambda, x_\mu) \to x_\lambda$, hence the result is a disc bundle over the manifold $\Sigma = D_1^\lambda \cup_\sim D_2^\lambda$, $x_\lambda \sim h(x_\lambda/|x_\lambda|)(1 - x_\lambda^2)^{1/2}$, the zero section of the bundle, cf. [S, § 18]. Since Σ is diffeomorphic to $\Sigma(h)$ (cf. 5.2), it is homeomorphic to a sphere.   □

The boundary of B is a $(\mu - 1)$-sphere bundle fibered by images in ∂B of the spheres $\{x_\lambda\} \times \partial D^\mu$, $x_\lambda \in \mathring{D}_i^\lambda$, $i = 1, 2$. The sphere $S_1^{\lambda-1}$ bounds the disc $K_1(a_+)$ in ∂D_1. The interior of $K_1(a_+)$ intersects transversely the fibers $\{x_\lambda\} \times \partial D^\mu$, thus its image K_{1B} in ∂B intersects transversely the fibers of ∂B. Of course, the same is true for the sphere $S_2^{\lambda-1}$, the disc Int $K_2(a_+)$ and its image K_{2B}. We note this for future reference:

(6.4)   *The diffeomorphism $M \to B$ sends the discs Int $K_1(a_+)$ and Int $K_2(a_+)$ onto discs that intersect transversely the fibers of the fibration $\partial B \to \Sigma$ (i.e., are partial sections).*   □

(6.5) Corollary   *If M is obtained from D^m by attaching a λ-handle along $h(S^{\lambda-1})$, where h: $S^{\lambda-1} \to \partial D^m$ is an imbedding extending to an imbedding of D^λ, then M is a μ-disc bundle over S^λ.*

Proof   By III,3.6, h is isotopic to the identity map $S^{\lambda-1} \to S^{\lambda-1} \subset \partial D^m$.   □

The condition on h is certainly satisfied if $\lambda = 1$ and $m > 1$. Since the only orientable disc bundle over the circle is the product bundle we obtain:

(6.6) Corollary   *If $D^m \cup H^1$ is orientable, then it is diffeomorphic to $S^1 \times D^{m-1}$.*   □

7 Cancellation Lemma

We will show in the next chapter that every closed manifold can be built by starting with a disc and consecutively attaching handles. It will then be important to recognize situations in which different sequences of attachments produce the same result. We present in this section two results of this type. The first one concerns the order in which handles are attached.

(7.1) Proposition *If $M_1 = (M \cup H^\mu) \cup H^\lambda$ and $\lambda \le \mu$, then M can be obtained by first attaching H^λ and then H^μ.*

Proof Let $\Sigma^{\lambda-1} \subset \partial(M \cup H^\mu), \Sigma^{\mu-1} \subset \partial M$ be the attaching spheres of handles H^λ, H^μ respectively. The assumption $\lambda \le \mu$ implies that dim $\Sigma^{\lambda-1}$ + dim(belt sphere of H^μ) = $\lambda - 1 + m - \mu - 1 < m - 1 = \dim \partial(M \cup H^\mu)$. Thus by IV,2.4 there is an isotopy of $\partial(M \cup H^\mu)$—hence of $M \cup H^\mu$—that pushes $\Sigma^{\lambda-1}$ off the belt sphere of H^μ. But any subset of $\partial(M \cup H^\mu)$ which is disjoint from the belt sphere is in $\partial M - \Sigma^{\mu-1}$. In particular, after the isotopy $\Sigma^{\lambda-1}$ is in ∂M and is disjoint from $\Sigma^{\mu-1}$. Now handles H^λ and H^μ can be attached in any order. $\square$

The second, deeper, result describes the situation when one handle cancels another, that is, when an attachment of two handles of consecutive dimensions to M produces no change in M. It will turn out that this happens when the attaching sphere of the second handle intersects the belt sphere of the first handle transversely in one point, and we begin by studying this condition.

(7.2) Lemma *Let $M = M_1 \cup H^\lambda$ and suppose that there is in ∂M a submanifold N intersecting the belt sphere $S^{\mu-1}$ transversely in one point. Then the attaching sphere Σ of H^λ bounds in ∂M_1 a manifold N' that is diffeomorphic to N with the interior of a disc removed.*

Proof Let $x = N \cap S^{\mu-1}$ and let $K(x)$ be as in 6.2. Since $K(x) \pitchfork S^{\mu-1}$, there is by IV,1.7 an isotopy of ∂M that brings N to coincide with $K(x)$ in a neighborhood of x and does not move the belt sphere. This is extended over M, using the collar of ∂M, in the usual way. Thus we can assume that N coincides with $K(x)$ along a small disc K centered at x (see Fig. VI,7). (If a proper tubular neighborhood in M_1 was used to construct M, then one can assume that $K = K(x)$, as it is easy to see.)

Remove the interior of K from N: what remains is a manifold N_1 in ∂M_1. Its boundary, a λ-sphere, equals $h\alpha(\partial K)$, where h is the attaching map. Now, applying h to the configuration in 6.2, we see that $h\alpha(\partial K)$ is

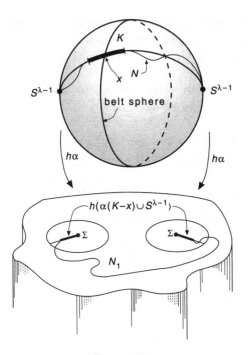

Figure VI,7

one of two components of the boundary of the manifold $h(\alpha(K - \{x\}) \cup S^{\lambda-1})$, diffeomorphic to $S^{\lambda-1} \times I$. The other component of the boundary is the attaching sphere $\Sigma = h(S^{\lambda-1})$. Thus

$$N' = N_1 \cup h(\alpha(K - \{x\}) \cup S^{\lambda-1})$$

is a manifold with boundary Σ and diffeomorphic to N_1.

Observe that Int N' is diffeomorphic to $N - \{x\}$ and that $N' \cap h(T)$ is a subbundle of $h(T)$: it equals $h(\alpha(K - \{x\}) \cup S^{\lambda-1})$, cf. 6.2.   □

(7.3) Theorem *Let $M = M_1 \cup H^\lambda$ and suppose that there is a λ-sphere S in ∂M that intersects the belt sphere of the handle transversely in one point x. Then there is a diffeomorphism of M onto $M_1 \#_b B$, where B is a disc bundle over a sphere, which maps S onto a section of this bundle.*

Proof Let N' be as in 7.2. Then N' is diffeomorphic to a closed disc, Int $N' = S - \{x\}$, and H^λ is attached along $\partial N'$. By 3.2 there is a

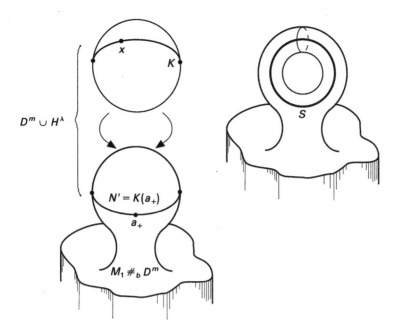

Figure VI,8

diffeomorphism $M_1 \to M_1 \#_b D^m$; by III,3.6 we may request it to map N' onto any preassigned disc in the boundary of D^m; we choose $K(a_+)$ as the image.

In this way we represented M as $M_1 \#_b (D^m \cup H^\lambda)$ where H^λ is attached to D^m along the boundary of $K(a_+)$, that is, along $S^{\lambda-1}$ (see Fig. VI,8). By 6.3, $D^m \cup H^\lambda$ is diffeomorphic to a disc bundle B.

In this representation of M, $S = \text{Int } N' \cup \{x\} = \text{Int } K(a_+) \cup \{x\}$. To see that the diffeomorphism $M \to B$ maps S onto a section, first note that by 6.4 it maps $K(a_+)$ onto a partial section. The only other intersection of S with fibers of B is the point x, and this intersection is transversal by assumption. Thus the image of S is a cross section by IV,1.3.2. □

The following theorem is the basic tool in the simplification of handle decompositions. It is sometimes referred to as the Cancellation Lemma and is due to S. Smale [Sm3].

(7.4)   Theorem   *Suppose that* $M = (M_1 \cup H^\lambda) \cup H^{\lambda+1}$, *where the attaching sphere of* $H^{\lambda+1}$ *intersects the belt sphere of* H^λ *transversely in one point. Then* M *is diffeomorphic to* M_1.

Proof   We take the attaching sphere of $H^{\lambda+1}$ as the sphere S in 7.3. Then, for some disc bundle B over a sphere,

$$M = (M_1 \#_b B) \cup H^{\lambda+1} = M_1 \#_b (B \cup H^{\lambda+1}) = M_1 \#_b D^m = M_1,$$

by successive application of 7.3, 5.3, and 3.2. $\square$

Exercise   Show that $P^m = D^m \cup H^1 \cup H^2 \cdots \cup H^m$. (*Hint*: Show that $P^m = B \cup H^m$, where B is a disc bundle over P^{m-1}; try the cases $m \le 3$ first.)

8   Combinatorial Attachment

The definition we have given of attaching handles presents a disadvantage in that M is not a subset of $M \cup H^\lambda$. This is inconvenient in homology computations. This problem would not occur if we defined $M \cup H^\lambda$ as $M \cup_h (D^\lambda \times D^\mu)$ where $h: \partial D^\lambda \times D^\mu \to \partial M$ is a diffeomorphism. This operation will be called the *combinatorial attaching of a handle*. It is often used to define handle attachment but it has a serious disadvantage in that it does not immediately yield a differentiable manifold. To obtain a smooth manifold one has to employ an additional procedure called *straightening the corners*. We will presently show that both definitions yield homeomorphic manifolds.

Assume that $M \cup H^\lambda$ is constructed using the imbedding h. Identify D^m with $D^\lambda \times D^\mu$ under a homeomorphism that sends $\partial D^\lambda \times D^\mu$ to the part of the boundary of D^m where $x_\lambda^2 \ge \frac{1}{2}$. Let h' be h restricted to the same part of the boundary.

(8.1)   Proposition $M \cup_{h'} D^m$ *is homeomorphic to* $M \cup H^\lambda$ *under a homeomorphism that is the identity on the belt disc and on the boundary of* $M \cup_{h'} D^m$.

Proof   Let $C = \{(x_\lambda, x_\mu) \in D^m \mid x_\lambda^2 \le \frac{1}{2}\}$ and $T_1 = \{(x_\lambda, x_\mu) \in D^m \mid x_\lambda^2 \ge \frac{1}{2}\}$. Then α interchanges $C - D^\mu$ and $T_1 - S^{\lambda-1}$ and is the identity on $C \cap T_1$. Hence the identification space $M \cup H^\lambda$ is identical with $(M - h(T_1)) \cup_{h_1} C$, $h_1 = h \mid C \cap T_1$.

Now expand C to cover the disc D^m; use the homeomorphism

$$g(x_\lambda, x_\mu) = (x_\lambda(2(1 - x_\mu^2))^{1/2}, x_\mu)$$

if $x_\mu^2 \leq \frac{1}{2}$ and the identity elsewhere in C. Similarly, expand $M - h(T_1)$ to cover M using the homeomorphism hgh^{-1} on $h(T)$ and the identity elsewhere. This yields a homeomorphism

$$(M - h(T_1)) \cup_{h_1} C \to M \cup_{g'} D^m,$$

where $g' = (h'gh^{-1})(hg^{-1}) = h'$, and proves the first part of the proposition. The second part follows from the fact that $g|D^\mu$ and $g|C \cap \partial D^m$ are identity maps. $\square$

Since $M \cup H^\lambda$ is a smooth manifold, 8.1 can be viewed as providing a smoothing procedure for $M \cup_h (D^\lambda \times D^\mu)$. Still another representation of $M \cup H^\lambda$ as $M \cup_h (D^\lambda \times D^\mu)$ can be obtained by shrinking C to $D^\lambda(1/2) \times D^\mu(1/2)$. In either representation $M \cup_h D^m$ contains $M \cup_h D^\lambda$. The subset of $M \cup_h D^m$ corresponding to D^λ is called the *core of the handle* (see Fig. VI,9). Clearly:

(8.2) Lemma $M \cup_h D^\lambda$ *is a strong deformation retract of* $M \cup_h D^m$. $\square$

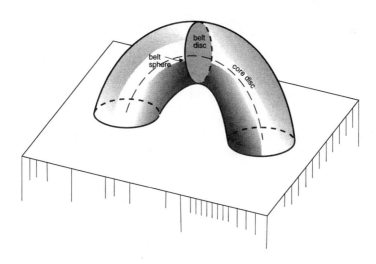

Figure VI,9

This implies easily:

(8.3)   Proposition   *The homomorphism $\pi_i(M) \to \pi_i(M \cup H^\lambda)$ is surjective if $\lambda > i$ and injective if $\lambda > i + 1$.*

Proof   By 8.2 we have to examine the inclusion of M into M with a λ-cell attached. But a map of S^i into $M \cup_h D^\lambda$ can be assumed to miss an interior point of D^λ if $\lambda > i$ and the same is true for a map of an $(i + 1)$-disc if $\lambda > i + 1$. $\square$

9   Surgery

Surgery on a $(\lambda - 1)$-sphere S in a manifold M^m is a special case of pasting: We paste M and S^m along S and $S^{\lambda-1}$. The resulting manifold will be denoted $\chi(M, S)$. With the notation of Section 6 it can be described as follows. Let $T' = \{x \in S^m \mid x_\lambda^2 > 0\}$; we view T' as a tubular neighborhood of $S^{\lambda-1}$ in S^m. Let $h: T' \to M$ be a diffeomorphism, $h(S^{\lambda-1}) = S$. Then $\chi(M, S) = (M - S) \cup_{h\alpha} (S^m - S^{\lambda-1})$, where α is as in 6.1.

Note that the operation of attaching a λ-handle along S becomes, when restricted to the boundaries, precisely surgery on S. This can be conveniently stated as follows. Consider h as an imbedding of T' in $M \times \{1\} \subset M \times I$ and attach a λ-handle to $M \times I$ along S. Let $W = (M \times I) \cup H^\lambda$; W is called the *trace of the surgery*.

(9.1)   Proposition   *If M is a closed manifold, then the boundary of W consists of $M \times \{0\}$ and $\chi(M \times \{1\}, S)$.* $\square$

We will refer to $M \times \{0\}$ and $\chi(M \times \{1\}, S)$ as the left- (resp. right-) hand boundaries of W and denote them $\partial_- W$ (resp. $\partial_+ W$). Observe that if we represent W as $(M \times I) \cup_h (D^\lambda \times D^\mu)$, as in 8.1, then the transversal disc D_t is represented by $0 \times D^\mu$ and the core disc D_c by $D^\lambda \times 0$, $\mu = \dim W - \lambda = m - \lambda + 1$. Now, $0 \times D^\mu \cup D^\lambda \times \partial D^\mu$ is a strong deformation retract of $D^\lambda \times D^\mu$. This implies immediately:

(9.2)   Lemma $\partial_+ W = \chi(M \times \{1\}, S)$ *with the transversal disc D_t attached along the belt sphere is a strong deformation retract of W.* $\square$

An argument analogous to that of 8.3 yields:

(9.3) Proposition *The homomorphism $\pi_i(\partial_+ W) \to \pi_i W$ is surjective if $i <$ $m - \lambda + 1$ and injective if $i < m - \lambda$.*   □

10   Homology and Intersections in a Handle

We will now prove some results concerning the effect of handle attachment on the homology of M.

The notation is that of 8.1, i.e., we represent W as $(M \times I) \cup_h (D_c \times D_t)$ with the transversal disc D_t of dimension $m - \lambda + 1$ and the core disc D_c of dimension λ. The boundary of D_c is the attaching sphere S and the boundary of D_t is the belt sphere S_β.

(10.1) Lemma (a) *The   inclusions $D_c \subset W$ and $D_t \subset W$ induce isomorphisms*

$$H_*(D_c, \partial D_c) \to H_*(W, M \times I),$$

$$H_*(D_t, \partial D_t) \to H_*(W, \partial_+ W).$$

(b) *The image of $\partial \colon H_\lambda(W, M \times I) \to H_{\lambda-1}(M \times I)$ is the subgroup $[S]$ generated by the fundamental class of S.*

Proof Part (a) follows from 8.2 and 9.2. Part (b) follows from the commutative diagram

$$
\begin{array}{ccc}
H_\lambda(W, M \times I) & \longrightarrow & H_{\lambda-1}(M \times I) \\
\uparrow{\scriptstyle\simeq} & & \uparrow \\
H_\lambda(D_c, \partial D_c) & \xrightarrow{\ \simeq\ } & H_{\lambda-1}(\partial D_c).
\end{array}
\qquad \square
$$

We will now study intersection numbers of submanifolds of W. For this we need a convention concerning orientations, and we assume the following.

(10.2) If W and D_c are oriented, then D_t is oriented so that $[D_c : D_t] = 1$.

In other words, the orientation of D_c followed by the orientation of D_t agrees with the orientation of W.

We denote the orientation of D_c, i.e., the chosen generator of $H_\lambda(D_c, \partial D_c)$, by g_c, and the orientation of D_t by g_t; the same letter will be used to denote the corresponding generators of $H_\lambda(W, M \times I)$ and $H_{m-\lambda+1}(W, \partial_+ W)$.

Suppose that there are given in the interior of W manifolds V_1 and V_2 representing homology classes $g_1 \in H_\lambda(W)$, $g_2 \in H_{m-\lambda+1}(W)$ and let $i_* : H_\lambda(W) \to H_\lambda(W, M \times I)$, $j_* : H_{m-\lambda+1}(W) \to H_{m-\lambda+1}(W, \partial_+ W)$ be induced by inclusions. Then $i_*(g_1) = \tau_1 g_c$ and $j_*(g_2) = \tau_2 g_t$, where $\tau_1 = [V_1 : D_t]$, by IV,5.1.

(10.3) Proposition $[V_1 : V_2] = \tau_1 \tau_2$.

Proof Consider the following diagram, in which all maps are induced by inclusion:

$$
\begin{array}{ccc}
H_{m-\lambda+1}(W) & \xrightarrow{\ \ k_*\ \ } & H_{m-\lambda+1}(W, W - V_1) \\
\Big\downarrow{\scriptstyle j_*} & & \Big\uparrow{\scriptstyle l_*} \\
H_{m-\lambda+1}(W, \partial_+ W) & \xrightarrow{\ \ =\ \ } & H_{m-\lambda+1}(D_t, \partial D_t).
\end{array}
$$

By IV,5.1, $l_*(g_t) = [D_t : V_1]g$ and $k_*(g_2) = [V_2 : V_1]g$, where g is an appropriately chosen generator of $H_{m-\lambda+1}(W, W - V_1)$. By commutativity

$$[V_2 : V_1]g = k_*(g_2) = l_* j_*(g_2) = \tau_2 l_*(g_t) = \tau_2 [D_t : V_1]g,$$

whence $[V_1 : V_2]g = \tau_1 \tau_2 g$, which proves 10.3. $\qquad\square$

In particular, it follows from 10.3 that if V represents a generator of $H_\lambda(W, M \times I)$, i.e., if $[V : D_t] = \pm 1$, then

(10.4) $j_*(g_2) = \pm [V_2 : V]g_t.$

One more simple relation will be needed in the future. Let S_α be an oriented λ-sphere in $\partial_+ W$ representing the class g_α in $H_\lambda(W)$; we want to identify its image $i_*(g_\alpha)$ in $H_\lambda(W, M \times I)$. Let S_α' be obtained by pushing S_α into the interior of W (using the collar of $\partial_+ W$). Then $i_*(g_\alpha) = [S_\alpha' : D_t]g_c$. But if $S_\beta = \partial D_t$ is the belt sphere oriented as the boundary of D_t, then $[S_\alpha' : D_t] = [S_\alpha : S_\beta]$ and we get

(10.5) $i_*(g_\alpha) = [S_\alpha : S_\beta]g_c.$

This relation holds without assuming W oriented; it is enough to orient D_c. This will yield g_c and a generator g_β of $H_\lambda(\partial_+ W, \partial_+ W - S_\beta)$, i.e., an orientation of the normal bundle to S_β.

11 (m, k)-Handlebodies, m > 2k

A manifold obtained by attaching g k-handles to the disc D^m is said to be an (m, k)-*handlebody of genus* g. To describe the structure of handlebodies we need the notion of a link. Let gD^k stand for a disjoint union of g copies of D^k. A *link* is an embedding h of the boundary $\partial(gD^k)$ in S^{m-1}. A link is trivial if h extends to an imbedding of gD^k. For instance, if $k = 1$ and $m > 2$, then every link is trivial. If an (m, k)-handlebody M is given by specified attaching maps then the restrictions of these maps to the attaching spheres determine a link, which we will call the *presentation link*. Observe that spheres of a presentation link have trivial normal bundles.

(11.1) Lemma *If the presentation link is trivial, then M is a boundary connected sum of g $(m - k)$-disc bundles over S^k.*

Proof Let h be the presentation link. Then $h = (h_1, \ldots, h_g)$, where each h_i is an imbedding of ∂D^k in ∂D^m extending to an imbedding $\bar{h}_i$ of D^k. Represent D^m as a connected sum along the boundary of g copies $D_1, \ldots, D_g$ of D^m. By III,3.7 we can assume that $\bar{h}_i$ sends D^k to D_i, and the lemma follows now from 6.5.   □

(11.2) Proposition *If $m \geq 2k + 1$, then an (m, k)-handlebody is a boundary connected sum of $(m - k)$-disc bundles over S^k.*

Proof We have to show that the presentation link of $(k - 1)$-spheres in S^{m-1} is trivial. If $m \geq 2k + 2$, then by II,3.2 each imbedding extends to an imbedding of a k-disc. We may assume that these discs are transversal; since $2k < \dim S^{m-1}$ they are then disjoint.

If $m = 2k + 1$, then we have a link of $(k - 1)$-spheres in S^{2k}. A theorem of Whitney quoted in II,4.7 asserts that each imbedding still extends to an imbedding of a k-disc and a transversality argument shows that we can assume this disc to be disjoint from all other spheres. To show that we can assume these discs to be disjoint from each other we proceed by induction. The inductive step is as follows. Suppose that we have in S^{2k} g imbedded k-discs disjoint from each other and from an imbedded $(k - 1)$-sphere S. Now, remove these discs from S^{2k}. By III,3.7, the resulting manifold is diffeomorphic to S^{2k} with g points removed; hence, by the same theorem of Whitney, S bounds a disc in it.   □

The only property of $\partial D^m = S^{m-1}$ that was used in proving 11.2 is that it is $(k-1)$-connected. Hence we have the following proposition, of which 11.2 is a special case.

(11.3) Proposition *Suppose that W is obtained by attaching g k-handles to $M \times I$ along $M \times \{1\}$. If M is $(k-1)$-connected and $\dim M \geq 2k$, then $W = (M \times I) \#_b T$, where T is a boundary connected sum of g disc bundles over S^k.* □

(11.4) Corollary *Assume $m > 2$ and let B_g be a $(m, 1)$-handlebody of genus g. Then:*

(*a*) *B_g is a connected sum along the boundary of g disc bundles;*
(*b*) *$B_g \#_b B_{g'} = B_{g+g'}$;*
(*c*) *Genus and orientability form a complete set of diffeomorphism invariants.*

Proof (a) follows from 11.2 and (b) follows from (a).

To prove (c) observe first that the presentation links of two $(m, 1)$-handlebodies of the same genus are isotopic. By 6.6, this implies (c) for

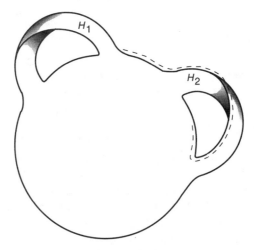

Dragging one leg of H_1 along the
marked path will untwist H_1

Figure VI,10

orientable handlebodies; in the non-orientable case one has to show that attaching two non-orientable handles is equivalent to attaching one orientable and one non-orientable handle. This is left as an exercise (drag one leg of a non-orientable handle across another non-orientable handle; see Fig. VI,10).   □

We will now calculate homology of handlebodies and their boundaries.

(11.5) Proposition *Let B be an* (m, k)*-handlebody of genus g. Then B has the homotopy type of a wedge of g k-spheres and its boundary is n-connected, where* $n = \min(k - 1, m - k - 2)$.

Proof By definition, B is obtained by attaching g k-handles to D^m along g disjoint $(k - 1)$-spheres $S_1, \ldots, S_q \subset D^m$. Let $S = \bigcup_i S_i$. By 8.2, D^m with g disjoint k-discs attached along S is a strong deformation retract of B. Following this deformation by a deformation of D^m to a point results in a wedge of g k-spheres.

The second statement is just a juxtaposition of 8.3 and 9.3.   □

Observe that if $m \geq 2k + 1$, then ∂B is $(k - 1)$-connected.

(11.6) Proposition *Let B be an* (m, k)*-handlebody of genus g. If* $m > 2k + 1$, *then* $H_k(\partial B)$ *is free abelian of rank g. If* $m = 2k + 1 > 2$, *then* $H_k(\partial B)$ *is free of rank 2g.*

Proof Let D stand for the union of all transversal $(m - k)$-discs of handles in B. We consider the part of the exact homology sequence of the pair $(\partial B \cup D, \partial B)$:

$$\cdots \to H_{k+1}(\partial B \cup D) \to H_{k+1}(\partial B \cup D, \partial B) \to H_k(\partial B)$$
$$\to H_k(\partial B \cup D) \to H_k(\partial B \cup D, \partial B) \to \cdots.$$

By 9.2 $\partial B \cup D$ is a strong deformation retract of B with an interior point removed; hence $H_i(\partial B \cup D) \simeq H_i(B)$ for $i < m - 1$. By 11.5 $H_k(B) = g\mathbf{Z}$ and is zero in all other positive dimensions. By excision $H_{m-k}(\partial B \cup D, \partial B) = H_{m-k}(D, \partial D) = g\mathbf{Z}$ and is zero in all other dimensions. Inserting these values in the exact sequence we obtain:

$$0 \to g\mathbf{Z} \to H_k(\partial B) \to g\mathbf{Z} \to 0 \qquad \text{if } m = 2k + 1 > 3;$$
$$0 \to 0 \to H_k(\partial B) \to g\mathbf{Z} \to 0 \qquad \text{if } m > 2k + 1. \quad \square$$

Exercise   Calculate the homology of the boundary of a $(3,1)$-handlebody.

It follows from 11.6 that two (m, k)-handlebodies with homeomorphic boundaries have the same genus if $m > 2k$ and that the boundary of such a handlebody is not a homotopy sphere unless $g = 0$. Both statements are definitely false if $m = 2k$. Removal of the interior of a disc from $S^k \times S^k$ leaves a $(2k, k)$-handlebody of genus 2 and a sphere as the boundary. This case is studied in the next section.

12 $(2k, k)$-Handlebodies; Plumbing

An imbedding h_i in the presentation link of a $(2k, k)$-handlebody B need not extend to an imbedding of a k-disc in ∂D^{2k}. However, according to a theorem of Whitney [Wi3], it does extend to an imbedding of a disc in D^{2k} if $k > 2$. This disc together with the core of the handle forms a k-dimensional sphere smoothly imbedded in the interior of B (8.1 is helpful in visualizing the situation here). In this way we obtain g imbedded spheres $\Sigma_1, \ldots, \Sigma_g$, which are oriented by the choice of orientations of cores of handles and are called *presentation spheres*. By 10.1(a) their fundamental classes yield a base for $H_k(B)$. (We could obtain the same result by starting with the base for $H_k(B)$ given by handles and then using the Hurewicz and Whitney theorems to realize it by imbedded spheres.) From now on we assume $k > 2$.

Now, the homology of $H_k(B, \partial B)$ has as a basis the transversal (belt) discs of handles $b_1, \ldots, b_g$. Let $v \in H_k(B)$ be represented by an imbedded oriented manifold V and let $j_*: H_k(B) \to H_k(B, \partial B)$ be induced by the inclusion. The following lemma is a consequence of 10.4.

(12.1)   Lemma $j_*(v) = \sum_j [V:\Sigma_j]b_j.$ $\square$

In other words, with the choice of bases are described, j_* is given by the matrix $\mathfrak{I} = ([\Sigma_i:\Sigma_j])$, $i, j = 1, \ldots, g$, which we will call the *intersection matrix* of the presentation. It is symmetric or skew-symmetric according to whether k is even or odd (see Fig. VI,11).

Consider now the exact homology sequence of the pair $(B, \partial B)$. Since the homology groups of B and of $(B, \partial B)$ vanish in all dimensions other than k, it follows that ∂B is $(k - 2)$-connected and $H_{k-1}(\partial B) \simeq$ Coker j_*, $H_k(\partial B) \simeq$ Ker j_*. Thus we obtain the following proposition.

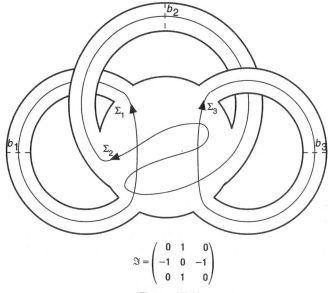

$$\mathfrak{S} = \begin{pmatrix} 0 & 1 & 0 \\ -1 & 0 & -1 \\ 0 & 1 & 0 \end{pmatrix}$$

Figure VI,11

(12.2) Proposition ∂B *is a homotopy sphere if and only if the intersection matrix* $\mathfrak{S}$ *is unimodular.*   □

Another consequence of 10.4 is that $[\Sigma_i : \Sigma_j]$ depends only on the relative position of the corresponding attaching spheres. That is, if S_i and S_j are two attaching spheres, $i \neq j$, s_i the fundamental class of S_i, and c_j a generator of $H_k(S^{2k-1} - S_j) \simeq \mathbf{Z}$, then

(12.3) $k_*(s_i) = \pm[\Sigma_i : \Sigma_j]c_j$,

where $k: S_i \hookrightarrow S^{2k-1} - S_j$ is the inclusion. The sign depends on the various choices of orientations. We leave the proof as an exercise. (Those familiar with the definition of linking numbers, e.g., [ST, § 77], will notice that this means that $[\Sigma_i : \Sigma_j]$ for $i \neq j$ is the linking number of S_i and S_j.)

12.3 provides an interpretation of the off-diagonal elements in $\mathfrak{S}$. The diagonal elements of $\mathfrak{S}$ are determined by the normal bundles of presentation spheres, as we will presently see.

Let $\phi_*: \pi_{k-1}(\mathbf{SO}(k)) \to \pi_{k-1}(S^{k-1})$ be the homomorphism induced by the projection in the fibration $\mathbf{SO}(k)/\mathbf{SO}(k-1) = S^{k-1}$ and let Σ be an imbedded k-sphere in a $2k$-dimensional manifold B, not necessarily a handlebody.

Then we have, by IV,5.4.1,

(12.4) $[\Sigma : \Sigma] = \phi_*(\alpha),$

where α is the characteristic element of the normal bundle of Σ, and we have identified $\pi_{k-1}(S^{k-1})$ with the group of integers.

Let $S_1, \ldots, S_g$ be g $(k-1)$-dimensional disjoint spheres with trivial normal bundles in ∂D^{2k} and let $\alpha_1, \ldots, \alpha_g$ be g elements of $\pi_{k-1}(\mathbf{SO}(k))$.

(12.5) **Proposition**   *There is a handlebody B such that the S_i are the attaching spheres and the α_i are the characteristic elements of the normal bundles of presentation spheres.*

Proof   The sphere S_i bounds a disc in D^{2k} and the tubular neighbourhood of the corresponding presentation sphere is obtained by attaching a handle to the tubular neighbourhood of this disc in D^{2k}; by 6.3.2 the attaching map can always be chosen so that the resulting disc bundle has α_i as characteristic element. $\square$

A theorem of S. Smale [Sm4,4.1] asserts that off-diagonal elements of every $g \times g$ symmetric or skew-symmetric matrix can be realized as linking numbers of a unique—up to isotopy—system of g $(k-1)$-dimensional disjoint spheres with trivial normal bundles in D^{2k}. This implies uniqueness in 12.5. We will not use it. Instead, we will construct explicitly a few examples with interesting properties.

Let τ_k be the characteristic element of the tangent bundle to the k-sphere. Then $\phi_*(\tau_k) = 0$ if k is odd and $=2$ if k is even; see A,5 for all necessary information about $\pi_{k-1}(\mathbf{SO}(k))$.

The first example is obtained by taking as S_1 and S_2 the intersections of ∂D^{2k} with, respectively, the subspace of the first k coordinates and the subspace of the last k coordinates. We let α_1 and α_2 be both equal τ_k. If k is odd the resulting handlebody $K(2k)$ has the intersection matrix

$$\begin{pmatrix} 0 & 1 \\ -1 & 0 \end{pmatrix}.$$

This is unimodular; thus $\partial K(2k)$ is a homotopy sphere. It is called the *Kervaire sphere*. If $k = 1, 3, 7$, then it is diffeomorphic to S^{2k-1}; this follows from the fact that the tangent bundle to S^k is trivial in this case; hence $K(2k)$ is diffeomorphic to $S^k \times S^k$ with a disc removed. The proof of this is left as an exercise.

If $k \neq 1, 3, 7$, then the situation is quite different: Kervaire proved that $\partial K(10)$ is not diffeomorphic to S^9 [K2]. In fact, according to W. Browder [Br2], $\partial K(2k)$ is not diffeomorphic to S^{2k-1} unless $k = 2^i - 1$, and it is diffeomorphic if $k = 15$. We say more about this in X,6.

Our construction can be generalized. We begin with g elements $\alpha_1, \ldots, \alpha_g$ of $\pi_{k-1}(\mathbf{SO}(k))$ and g points "weighted" by $\alpha_1, \ldots, \alpha_g$. Parametrize D^{2k} as $D^k \times D^k$ and attach a k-handle, another $D^k \times D^k$, by the map

$$(x, y) \mapsto (x, \alpha_1(x) \cdot y), \qquad x \in \partial D^k, \qquad y \in D^k.$$

The result is a disc bundle over S^k with characteristic element α_1 and zero section Σ_1. Retain the parameterization in the handle and attach to it another k-handle along the boundary of the belt disc $\mathbf{0} \times D^k$ using the map

$$(x, y) \mapsto (\alpha_2(y) \cdot x, y)), \qquad x \in D^k, \qquad y \in \partial D^k.$$

At the same time join the point weighted by α_1 to the point weighted by α_2. Note that the first handle with the second handle attached to it is again a disc bundle with characteristic element α_2 and the zero section $\Sigma_2 = $ (belt disc) $\cup$ (core of the handle). Clearly $[\Sigma_1 : \Sigma_2] = \pm 1$.

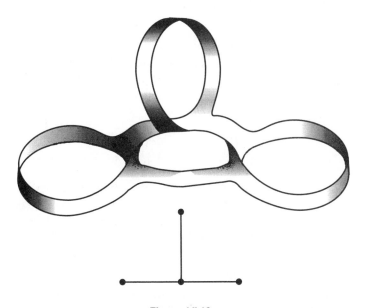

Figure VI,12

This attachment of handles is continued, but we have to decide to which handle, first or second, the third handle is to be attached. Once the choice is made, we attach it to the boundary of a belt disc of the chosen handle using α_3, and join the third point to the first or second according to the choice. This creates the new sphere Σ_3 with intersection numbers either $[\Sigma_3 : \Sigma_2] = \pm 1$ or $[\Sigma_3 : \Sigma_1] = \pm 1$. The result is a handlebody, for we attached a k-handle to the boundary of a handlebody.

The end result of this construction is a $(2k, k)$-handlebody with presentation spheres $\Sigma_1, \ldots, \Sigma_g$, each with the normal bundle α_i, and a weighted graph that completely describes it (see Fig. VI,12). An obvious induction argument shows that this graph is contractible: It is a tree.

To the Kervaire manifold $K(4n + 2)$ there corresponds the graph with vertices weighted by τ_{2n+1}. To the graph

with all vertices weighted by τ_{2n} there corresponds a $4n$-dimensional handlebody $M(4n)$ with the intersection matrix

$$\Gamma_8 = \begin{pmatrix} 2 & 1 & & & & & & \\ 1 & 2 & 1 & & & & 0 & \\ & 1 & 2 & 1 & & & & \\ & & 1 & 2 & 1 & & & \\ & & & 1 & 2 & 1 & 1 & 0 \\ & & & & 1 & 2 & 0 & 0 \\ & & 0 & & & 1 & 0 & 2 & 1 \\ & & & & & 0 & 0 & 1 & 2 \end{pmatrix}$$

Since Γ_8 is unimodular, $\partial M(4n)$ is a homotopy sphere. A calculation shows that the quadratic form over the reals with the matrix Γ_8 has signature 8. We will show in IX,8 that $\partial M(4n)$ is not diffeomorphic to S^{4n-1} for $n = 2,3$.

This construction we just described is called *plumbing of disc bundles*. Each step in it results in augmenting the intersection matrix by a column with the following property:

(*)   All elements above the diagonal are zero with the exception of one that equals ± 1.

Of course, the elements on the diagonal are determined by 12.4, and the elements below the diagonal by the symmetry or skew-symmetry of the matrix. It is clear that all matrices in which columns satisfy (∗) can be obtained. There is a modification of plumbing due to W. Browder that yields all matrices with even elements on the diagonal [Br1,V]. This follows also from the just quoted theorem of Smale, but Browder's construction is explicit. For another description of plumbing see [Hr].

VII

Handle Presentation Theorem

The handle presentation theorem of Milnor and Wallace asserts that every manifold can be constructed by successive attachment of handles. We state it here in terms of elementary cobordisms. This and related notions are introduced in Section 1; the theorem itself is proved in Section 2.

In Section 3 we show how to calculate the homology of a manifold from its handle presentation; this is applied in Section 4 to deduce the Morse inequalities. In Section 5 we discuss handle presentations of oriented manifolds and derive a version of the Poincaré duality theorem for manifolds with boundary. In Section 6 we show how to obtain a handle presentation with a minimal number of 0-handles. A classical application to 3-manifolds (Heegaard diagram) follows in Section 7.

The handle presentation theorem is the starting point of the proof of the h-cobordism theorem presented in the next chapter.

1 Elementary Cobordisms

An ordered triple of manifolds $\mathscr{C} = \{V_0, W, V_1\}$ is called a *cobordism* if $\partial W = V_0 \cup V_1$ and V_0, V_1 are disjoint open subsets of ∂W. We will often write $V_0 = \partial_- W$, $V_1 = \partial_+ W$ and call V_0 (resp. V_1) the left-hand (resp. right-hand) boundary of W. Throughout this chapter we will consider only cobordisms with W compact.

The simplest example of a cobordism is the trivial cobordism $\{M \times \{0\}, M \times I, M \times \{1\}\}$ where M is a compact closed manifold. We have encountered another example of a cobordism in VI,9.1: $\{M \times \{0\}, M \times I \cup H^\lambda, \chi(M \times \{1\}, S)\}$. This will be called an *elementary cobordism* of index λ; it is the result of attaching a λ-handle to the right-hand boundary of $M \times I$.

It will be convenient to view a trivial cobordism as an elementary cobordism of index -1.

Suppose we are given two cobordisms $\mathscr{C} = \{V_0, W, V_1\}$, $\mathscr{C}' = \{V_0', W', V_1'\}$ and a diffeomorphism $h\colon V_1 \to V_0'$. As in VI,5 we can join W and W' using h; let $W_1 = W \cup_h W'$. Then $\partial W_1 = V_0 \cup V_1'$ and $\{V_0, W_1, V_1'\}$ is a cobordism, which will be denoted $\mathscr{C} \cup \mathscr{C}'$. Again, this is a symbolic notation in that it does not show the diffeomorphism h on which the result depends. However, if $\mathscr{C}'$ is a trivial cobordism then the result does not depend on h: by VI,5.3 we then have $\mathscr{C} \cup \mathscr{C}' = \mathscr{C}$.

The fundamental role played by elementary cobordisms is explained by the following theorem due to Smale and Wallace. Since elementary cobordisms amount to attaching a handle it will be called the Handle Presentation Theorem.

(1.1) Theorem *Let $\mathscr{C}$ be a cobordism. Then $\mathscr{C} = \mathscr{C}_1 \cup \mathscr{C}_2 \cup \cdots \cup \mathscr{C}_k$, where the $\mathscr{C}_i$ are elementary cobordisms. Moreover, one can assume that $i < j$ implies $\lambda(i) \le \lambda(j)$, where $\lambda(i)$ denotes the index of $\mathscr{C}_i$.*

The proof will be given following the proof of 2.2 in the next section.

(1.2) Corollary *Let $\mathscr{C} = \{V_0, W, V_1\}$ be a cobordism. Then there exists a sequence of manifolds $V_0 \times I = W_{-1} \subset W_0 \subset W_1 \subset \cdots \subset W_m = W$ such that W_i is obtained from W_{i-1} by attaching a number of i-handles to its right-hand boundary.*

Proof Represent $\mathscr{C}$ as in 1.1. Let W_i be the union of all cobordisms of index $\le i$ in this presentation. Then W_{i+1} is obtained from W_i by attaching in succession a certain number of $(i+1)$-handles. Since they are all of the same index they can all be attached "at the same time," i.e., W_{i+1} can be obtained from W_i by attaching a certain number of $(i+1)$-handles to its right-hand boundary (*cf.* VI,7.1).   □

The sequence $\{W_i\}$ of manifolds will be called the *presentation* of $\mathscr{C}$, the manifold W_i its ith level.

In the future it will be sometimes convenient to represent W_i as

$$W_{i-1} \cup (\partial_+ W_{i-1} \times I \cup H_1^i \cup H_2^i \cup \cdots \cup H_k^i).$$

2 Handle Presentation Theorem

The handle presentation in 1.1 will be shown to be a simple consequence of the existence of a Morse function. The link between cobordisms and the theory of Morse functions is provided by 2.1 and 2.2.

Let M be a compact manifold with or without boundary and let $f : M \to \mathbf{R}$ be a smooth function. We set $M_a = f^{-1}(-\infty, a]$, $M_{a,b} = f^{-1}[a, b]$. If a and b are regular values of f, $a < b$, and $M_{a,b} \cap \partial M = \emptyset$, then, by II,2.5, $M_{a,b}$ is a manifold with boundary $f^{-1}(a) \cup f^{-1}(b)$. In particular, we have a cobordism $\mathscr{C} = \{f^{-1}(a), M_{a,b}, f^{-1}(b)\}$.

(2.1) Proposition *If f has no critical points in $M_{a,b}$, then $\mathscr{C}$ is a trivial cobordism.*

Proof This is just a restatement of I,7.5.

(2.2) Proposition *If f has exactly one critical point in $M_{a,b}$ and it is of index λ, then $\mathscr{C}$ is an elementary cobordism of index λ.*

Proof Let p be the critical point of f in $M_{a,b}$; we assume $f(p) = 0$. By IV,4.2 there is a chart U at p, which we will simply identify with $\mathbf{R}^m = \mathbf{R}^\lambda \times \mathbf{R}^\mu$, such that $p = \mathbf{0}$ and $f(x) = -x_\lambda^2 + x_\mu^2$ in some neighborhood U, say $x^2 < 100$, of $\mathbf{0}$. (x_λ, x_μ are projections of x into $\mathbf{R}^\lambda, \mathbf{R}^\mu$.)

Let $\varepsilon > 0$ be such that $a < -\varepsilon$, $\varepsilon < b$. Then f has no critical points in either $M_{a,-\varepsilon}$ or $M_{\varepsilon,b}$. By 2.1 these two are trivial cobordisms, and to prove 2.2 we have to show that

(2.2.1) M_ε is diffeomorphic to $M_{-\varepsilon}$ with a λ-handle attached.

This will be done in two steps. The "difference" between M_ε and $M_{-\varepsilon}$ is a large manifold extending beyond U. In the first step we will use an argument due to J. Milnor [M1] to find another manifold contained in and diffeomorphic to M_ε, but such that the difference between it and $M_{-\varepsilon}$ is contained in U. This will enable us in the second step to use the coordinate system in U to construct the required diffeomorphism explicitly.

We will need a non-negative function $\phi(t)$ such that

(2.2.2) $\phi(0) > \varepsilon,$ $\phi(t) = 0$ for $t \geq 2\varepsilon,$ $-1 < \phi'(t) \leq 0.$

This is easy to construct. Given ϕ, we set

$$F(x) = \begin{cases} f(x) - \phi(x_\lambda^2 + 2x_\mu^2) & \text{for x in U,} \\ f(x) & \text{elsewhere.} \end{cases}$$

We now prove that

(2.2.3) M_ε is diffeomorphic to $F^{-1}(-\infty, -\varepsilon]$.

To see this, note first that $M_\varepsilon = F^{-1}(-\infty, \varepsilon]$. For, clearly, $M_\varepsilon \subset F^{-1}(-\infty, \varepsilon]$. On the other hand, if $x \in F^{-1}(-\infty, \varepsilon]$ and $\phi(x_\lambda^2 + 2x_\mu^2) > 0$, then $x_\lambda^2 + 2x_\mu^2 < 2\varepsilon$; hence $f(x) = -x_\lambda^2 + x_\mu^2 \leq \tfrac{1}{2}x_\lambda^2 + x_\mu^2 < \varepsilon$, i.e., $x \in M_\varepsilon$.

Now a simple computation shows that $\nabla F = 0$ if and only if $\nabla f = 0$, that is, that F has the same critical points as f. Since $F(p) < -\varepsilon$, F has no critical points in $F^{-1}[-\varepsilon, \varepsilon]$; thus, by 2.1, $F^{-1}(-\infty, \varepsilon]$ is diffeomorphic to $F^{-1}(-\infty, -\varepsilon]$. This proves 2.2.3.

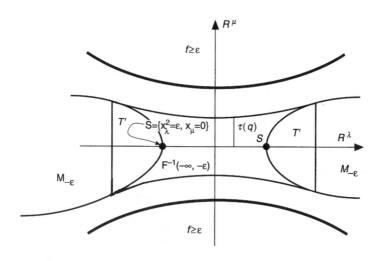

Figure VII,1

We have accomplished the first step: The difference between $M_{-\varepsilon}$ and $F^{-1}(-\infty, -\varepsilon]$ is contained in U. Before we tackle the second step we need a computation.

(2.2.4) Let $q \in \mathbf{R}^\lambda$. The intersection of $F^{-1}(-\infty, -\varepsilon]$ with the μ-plane $x_\lambda = q$ is diffeomorphic to a disc of radius $r(q) > 0$. The function $r(q)$ is smooth and if $q^2 > 2\varepsilon$, then $r(q) = (q^2 - \varepsilon)^{1/2}$. (See Fig. VII,1.)

The intersection in question consists of points $x = (x_\lambda, x_\mu)$ satisfying

$$(*) \qquad x_\lambda = q, \qquad -q^2 + x_\mu^2 - \phi(q^2 + 2x_\mu^2) \le -\varepsilon.$$

With $t = q^2 + 2x_\mu^2$ this becomes

$$\phi(t) \ge \frac{t}{2} + \varepsilon - \frac{3}{2}q^2,$$

which, since $\phi(t) - t/2$ is monotone decreasing, is satisfied for all $t \le t_0$, where $\phi(t_0) = t_0/2 + \varepsilon - (3/2)q^2$. This is the same as saying that $(*)$ is satisfied by all points x_μ such that $x_\mu^2 \le \frac{1}{2}(t_0 - q^2)$. Now, by 2.2.2, $\phi(t) - \phi(0) > -t$; hence

$$t_0/2 + \varepsilon - \frac{3}{2}q^2 = \phi(t_0) > \phi(0) - t_0 > \varepsilon - t_0,$$

i.e., $t_0 > q^2$. This proves 2.2.4 with $r(q) = (\frac{1}{2}(t_0 - q^2))^{1/2}$. Since t_0 is a smooth function of q, $r(q)$ is smooth.

The proof of 2.2 will now be concluded by showing that

(2.2.5) $F^{-1}(-\infty, -\varepsilon]$ is diffeomorphic to $M_{-\varepsilon}$ with a λ-handle attached.

This λ-handle is attached to $M_{-\varepsilon}$ along the $(\lambda - 1)$-sphere $S = \{x_\mu = 0, x_\lambda^2 = \varepsilon\}$ as in VI,6.1; the attaching map is the diffeomorphism

$$h(x_\lambda, x_\mu) = \sqrt{2\varepsilon}\left(\frac{x_\lambda}{|x_\lambda|}(3/2 - x_\lambda^2)^{1/2}, x_\mu\right)$$

of the tubular neighborhood $T(\varepsilon) = \{x \in D^m \,|\, x_\lambda^2 > \varepsilon\}$ of $S^{\lambda-1}$ onto a tubular neighborhood $T' = \{x \in M_{-\varepsilon} \,|\, x_\lambda^2 < 2\varepsilon\}$ of S in $M_{-\varepsilon}$.

Thus $M_{-\varepsilon} \cup H^\lambda$ is $(M_{-\varepsilon} - S) \cup_\sim (D^m - S^{\lambda-1})$, with the identification

$$\textbf{(2.2.6)} \qquad (x_\lambda, x_\mu) \sim h\alpha(x_\lambda, x_\mu) = \sqrt{2\varepsilon}\left(x_\lambda, x_\mu\left(\frac{x_\lambda^2 - 1/2}{1 - x_\lambda^2}\right)^{1/2}\right),$$

where α is as in VI,6.1.

Now the diffeomorphism $g\colon M_{-\varepsilon} \cup H^\lambda \to F^{-1}(-\infty, -\varepsilon]$ is given by:

$$(2.2.7) \qquad g(x) = \begin{cases} \sigma(x, x_\mu) = \left(x_\lambda, x_\mu \dfrac{r(x_\lambda)}{\sqrt{x_\lambda^2 - \varepsilon}}\right) & \text{if } x \in M_{-\varepsilon} - S, \\[3mm] \tau(x, x_\mu) = \left(x_\lambda \sqrt{2\varepsilon}, x_\mu \dfrac{r(x_\lambda \sqrt{2\varepsilon})}{\sqrt{1 - x_\lambda^2}}\right) & \text{if } x \in D^m - S^{\lambda-1}. \end{cases}$$

By 2.2.4 σ and τ are both smooth, and since $\sigma h\alpha = \tau$, g is a well-defined smooth map; its inverse is easily calculated, showing that g is a diffeomorphism.

To compute the image of g, note first that the intersection of $M_{-\varepsilon} - S$ with a plane $x_\lambda = q$ is a disc of radius $(q^2 - \varepsilon)^{1/2}$ and that σ maps this disc onto the intersection of $F^{-1}(-\infty, -\varepsilon]$ with $x_\lambda = q$. Hence $\sigma(M_{-\varepsilon} - S) = F^{-1}(-\infty, -\varepsilon] \cap \{(x_\lambda, x_\mu) \mid x_\lambda^2 > \varepsilon\}$.

Similarly, τ maps the intersection of $D^m - S^{\lambda-1}$ with the plane $x_\lambda = q$ onto the intersection of $F^{-1}(-\infty, -\varepsilon]$ with $x_\lambda = \sqrt{2\varepsilon}\, q$. Thus $\tau(D^m - S^{\lambda-1}) = F^{-1}(-\infty, -\varepsilon] \cap \{(x_\lambda, x_\mu) \mid x_\lambda^2 < 2\varepsilon\}$ and, finally, $g(M_{-\varepsilon} \cup H^\lambda) = \sigma(M_{-\varepsilon} - S) \cup \tau(D^m - S^{\lambda-1}) = F^{-1}(-\infty, -\varepsilon]$. $\square$

We can prove 1.1 now. If $\{V_0, W, V_1\}$ is a cobordism, then, by IV,3.5, there is a Morse function on W taking the value 0 on V_0, 1 on V_1, and having distinct critical values. Thus the first part of 1.1 follows from 2.2 and the second part from VI,7.1.

We have shown that a Morse function on a cobordism yields a presentation. Conversely, to every presentation there corresponds a Morse function that yields it. To see this, it is enough to show that for every elementary cobordism of index λ there is a Morse function constant on the boundary and with only one critical point of index λ. We will show this by adapting the construction used in the proof of 2.2.

Let $W = V \times [-2\varepsilon, -\varepsilon]$ and suppose that a handle H^λ is attached to $\partial_+ W$ along a sphere Σ. Let T be a tubular neighborhood of Σ in W. Let d be a diffeomorphism of T onto the tubular neighborhood $T' = \{x \in M_{-\varepsilon} \mid x_\lambda^2 < 2\varepsilon\}$ of S in $M_{-\varepsilon}$; d can be chosen so that $fd(x, t) = t$, where $f(x_\lambda, x_\mu) = -x_\lambda^2 + x_\mu^2$, $(x, t) \in W$.

We will identify T with T'; we thus can view the handle H^λ as attached to W along T' and our task is to extend the function $(x, t) \mapsto t$ over the handle.

Now, if H^λ is attached via the identification 2.2.6, then the extension, and the desired function with only one critical point of index λ, is simply given by setting

$$G(q) = \begin{cases} t & \text{if } q = (x, t) \in W - T, \\ Fg \text{ where } g \text{ is given by 2.2.7} & \text{if } q \in D^m - S^{\lambda-1}, \end{cases}$$

In the general case H^λ is attached by an identification that is a composition of 2.2.6 with a rotation in the μ-coordinate. Since F is invariant under such a rotation, the preceding formula for G works in the general case as well.

It follows from this argument that if there is given a presentation $\mathscr{P}$ of a cobordism $\{V_0, W, V_1\}$, then $\mathscr{P}$ is derived from some Morse function f on W. In this case the function $-f$ yields a presentation of the cobordism $\{V_1, W, V_0\}$, called the *dual presentation*.

(2.3) Proposition *To every presentation $\mathscr{P}$ there corresponds a dual presentation $\bar{\mathscr{P}}$ such that the λ-handles of $\mathscr{P}$ are the $(m - \lambda)$-handles of $\bar{\mathscr{P}}$, and such that the attaching sphere of a λ-handle of $\mathscr{P}$ is the belt sphere of the corresponding $(m - \lambda)$-handle of $\bar{\mathscr{P}}$ and vice versa.*

Proof With the same assumptions as in 2.2, let $\mathscr{C}'$ be the cobordism $\{f^{-1}(b), M_{a,b}, f^{-1}(a)\}$. Consideration of the function $-f + a + b$ shows that $\mathscr{C}'$ is an elementary cobordism of index $m - \lambda$, and that the attaching sphere of $\mathscr{C}'$ is the belt sphere of $\mathscr{C}$ and vice versa. □

This procedure is sometimes called "turning the cobordism upside down." There is a good reason (*cf.* 5.1 following)—besides brevity—to use the word *dual* instead.

3 Homology Data of a Cobordism

The presentation of a cobordism $\{V_0, W, V_1\}$, as in 1.2, allows us to read homology properties of W in a much simpler and more geometric way than a triangulation does. This is so because this filtration is cellular ([D,VI.1]) and the composition of homomorphisms

(3.1) $H_k(W_k, W_{k-1}) \to H_{k-1}(W_{k-1}) \to H_{k-1}(W_{k-1}, W_{k-2})$

has a very simple geometric interpretation in terms of intersection numbers. This can be described precisely as follows.

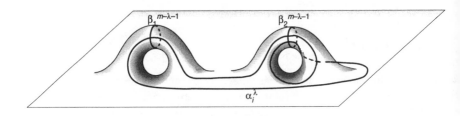

Figure VII,2

Let $\mathscr{C} = \{V_0, W, V_1\}$ be a cobordism and $\mathscr{P}$ its presentation as in 1.2. Fix an orientation for the core of every λ-handle H_i^λ, $\lambda = 0, \ldots, m$, $i = 1, \ldots, c_\lambda$. This will induce an orientation of the normal bundle to the belt sphere $\beta_i^{m-\lambda-1}$ and, via the attaching map, of the attaching sphere $\alpha_i^{\lambda-1}$ (see Fig. VII,2). Thus all intersection numbers $[\alpha_i^{\lambda-1} : \beta_j^{m-\lambda}]$ are now defined as in VI,10.5. For a given λ, $1 \le \lambda \le m$, we will arrange these in a $(c_{\lambda-1}) \times c_\lambda$ matrix $\mathfrak{M}_\lambda$ in which rows are indexed by $(\lambda - 1)$-handles, columns by λ-handles and $[\alpha_i^{\lambda-1} : \beta_j^{m-\lambda}]$ stands at the intersection of the ith column and the jth row. Finally, let C_λ be the free abelian group generated by λ-handles $H_1^\lambda, \ldots, H_{c_\lambda}^\lambda$. If $c_\lambda = 0$ we let C_λ be the trivial group. In this case we let $\mathfrak{M}_\lambda$ and $\mathfrak{M}_{\lambda+1}$ be trivial matrices, that is, to have all entries equal zero.

It is clear that if $\mathscr{P}$ is obtained from a Morse function f, then c_λ is the number of critical points of f of index λ.

It will be convenient to represent an element $v = \sum m_i H_i^\lambda$ of C_λ as a $c_\lambda \times 1$ matrix $'(m_1, \ldots, m_{c_\lambda})$. With this notation we define a homomorphism $\partial_\lambda : C_\lambda \to C_{\lambda-1}$ by

$$(3.2) \qquad\qquad \partial_\lambda v = \mathfrak{M}_\lambda \cdot v,$$

i.e., $\partial H_i^\lambda = \sum_j [\alpha_i^{\lambda-1} : \beta_j^{m-\lambda}] H_j^{\lambda-1}$.

(3.3) Definition The graded group $\{C_\lambda\}$ and the set of matrices $\mathfrak{M}_\lambda$ will be called the *homology data* of $\mathscr{P}$.

(3.4) Theorem $C_* = \{C_\lambda, \partial_\lambda\}$ is a chain complex, and $H_*(C_*) = H_*(W, V_0)$.

Proof Assign to a λ-handle the class in $H_\lambda(W_\lambda, W_{\lambda-1})$ corresponding to the orientation of its core. By VI,10.1(a), this induces an isomorphism

$C_\lambda \to H_\lambda(W_\lambda, W_{\lambda-1})$. By VI,10.1(b) and VI,10.5 this isomorphism makes ∂_λ correspond to the composition 3.1. Thus the theorem follows from [D,VI.3]. □

We have shown how the homology data of a presentation of a cobordism W determine the homology of W with integral coefficients. However, if R is any principal ideal domain and we let C_λ be the free R-module generated by the set of λ-handles, then VI,10.5 will still make sense, hence 3.2 as well, and the proof of 3.4 (with the reference changed to [D,VI,7.11]) will remain valid for homology with coefficients in R.

If $R = \mathbf{Z}/2\mathbf{Z}$, then the theory is considerably simplified by the fact that intersection numbers can be replaced by intersection numbers mod 2, i.e., the number mod 2 of points of intersection, and no orientations have to be chosen.

Exercise Represent P^2 as in 1.2: $D^2 \subset W_1 \subset P^2$, where W_1 is a Möbius strip, and calculate the integral and mod 2 homology of P^2 from this presentation using 3.4.

Exercise Let p, q be a pair of relatively prime integers. A (p, q)-torus knot is a simple closed curve S on the surface of the solid torus $T = D^2 \times S^2$ that wraps p times in the longitudinal direction and q times in the meridianal direction. Attach a 2-handle H^2 to T along S and show that the boundary of $T \cup H^2$ is a 2-sphere. (Use combinatorial handles as in VI,8 and prove first that a surface of T with a (p, q)-torus knot removed is an annulus.)

Attach a 3-handle to $T \cup H^2$ along the boundary. The result is a closed, compact 3-dimensional manifold called the *lens space* $L(p, q)$. Since $T = D^3 \cup H^1$ we have a presentation $L(p, q) = D^3 \cup H^1 \cup H^2 \cup H^3$. Calculate the homology of $L(p, q)$.

Exercise Let $E(n)$ be the total space of the disc bundle associated to the Hopf fibration $S^{n-1} \to S^{n/2}$, $n = 4, 8, 16$. Attaching an n-disc (n-handle) to $E(n)$ along the boundary produces a compact manifold $M(n)$. Find a presentation and the homology data of $M(n)$.

It is worthwhile to emphasize that all constructions in this section were made without any assumption on the global orientability of W; only orientations of cores were used, and these were chosen arbitrarily. In particular, the orientations of cores of $\mathscr{P}$ do not induce any orientation of the dual presentation $\bar{\mathcal{P}}$. However, if W is an oriented manifold, then a preferred way to orient $\bar{\mathcal{P}}$ is given as follows. Denote the oriented core of the handle H_i^λ of $\mathcal{P}$ by c_i^λ and orient the transverse disc $t_i^{m-\lambda}$ by the convention VI,10.2; this can be written symbolically

$$(*) \qquad\qquad c_i^\lambda \wedge t_i^{m-\lambda} = W,$$

where " $\wedge$ " means "followed by" and the equality sign applies to orientations.

Now, the cores of $\bar{\mathcal{P}}$ are the transverse discs of $\mathcal{P}$. We extend this rule to their orientations, that is, we orient the cores of $\bar{\mathcal{P}}$ by the rule

$$(**) \qquad\qquad \bar{c}_i^{m-\lambda} = t_i^{m-\lambda}.$$

$\bar{\mathcal{P}}$ is now oriented; we call the resulting homology data $\{\bar{C}_\lambda\}, \bar{\mathfrak{M}}_\lambda$ the *dual homology data*. Note that $\bar{C}_\lambda$ is generated by the λ-handles of $\bar{\mathcal{P}}$, that is, by the $(m-\lambda)$-handles of $\mathcal{P}$. The relation between the intersection matrices is given by:

(3.5)   Proposition $(-1)^{m-\lambda+1}\mathfrak{M}_\lambda = {}^t\bar{\mathfrak{M}}_{m-\lambda+1}.$

Proof The key to the proof is that in an oriented manifold the intersection numbers can be expressed as intersection numbers of oriented submanifolds, provided that the transversal discs are oriented by VI,10.2. For $\mathcal{P}$ this has already been done in $(*)$; for $\bar{P}$ we set analogously

$$\bar{c}_i^{m-\lambda} \wedge \bar{t}_i^\lambda = W.$$

Comparing this with $(*)$ we get

$$c_i^\lambda \wedge t_i^{m-\lambda} = \bar{c}_i^{m-\lambda} \wedge \bar{t}_i^\lambda = (-1)^{\lambda(m-\lambda)} \bar{t}_i^\lambda \wedge t_i^{m-\lambda};$$

hence $c_i^\lambda = (-1)^{\lambda(m-\lambda)} \bar{t}_i^\lambda$.

Passing to the boundaries in this and in $(**)$, we get

$$(***) \qquad \alpha_i^{\lambda-1} = (-1)^{\lambda(m-\lambda)} \bar{\beta}_i^{\lambda-1}, \qquad \bar{\alpha}_i^{m-\lambda-1} = \beta_i^{m-\lambda-1}.$$

Now, $\bar{\alpha}_i^{m-\lambda}$ and $\bar{\beta}_j^{\lambda-1}$ are submanifolds of $\partial_+\bar{W}_{m-\lambda}$; $\alpha_j^{\lambda-1}, \beta_i^{m-\lambda}$ are submanifolds of $\partial_+ W_{\lambda-1}$. Since $\partial_+\bar{W}_{m-\lambda}$ is the same manifold as $\partial_+ W_{\lambda-1}$ but with opposite orientation, we conclude from $(***)$ that

$$[\bar{\alpha}_i^{m-\lambda} : \bar{\beta}_j^{\lambda-1}] = -(-1)^{\lambda(m-\lambda)}[\beta_i^{m-\lambda} : \alpha_j^{\lambda-1}] = (-1)^{m-\lambda+1}[\alpha_j^{\lambda-1} : \beta_i^{m-\lambda}],$$

which is precisely 3.5. $\square$

4   Morse Inequalities

We will apply 3.4 to obtain the celebrated theorems of M. Morse.

(4.1) Theorem *Suppose that M^m is a compact, closed manifold and f a Morse function on M. Let c_i be the number of critical points of f of index i and let $b_i = \operatorname{rank} H_i(M)$ (the ith Betti number of M). Then, for every n,*

(4.1.n) $$b_n - b_{n-1} + b_{n-2} - \cdots \le c_n - c_{n-1} + c_{n-2} - \cdots.$$

Proof Recall ([D,V.5]) that if $C_* = \{C_i\}$ is a complex where all the C_i are of finite rank and almost all are of rank zero then

$$(*) \qquad\qquad b_0 - b_1 + b_2 - \cdots = c_0 - c_1 + c_2 - \cdots,$$

where $b_i = \operatorname{rank} H_i(C_*)$ and $c_i = \operatorname{rank} C_i$.

Now, the function f yields for the cobordism $\{\emptyset, M, \emptyset\}$ the homology data 3.3, i.e., the chain complex $C_* = \{C_i, \partial_i\}$ such that $H_*(C_*) = H_*(M)$ and rank $C_i = c_i = $ number of critical points of f of index i. For a given number n, $0 \le n < m$, consider also the chain complex $C_*^{(n)} = \{C_i, \partial_i\}_{i \le n}$. By $(*)$ we have

$$b_n' - b_{n-1}' + b_{n-2}' - \cdots = c_n - c_{n-1} + c_{n-2} - \cdots,$$

where $b' = \operatorname{rank} H_i(C_*^{(n)})$. Since $H_i(C_*^{(n)}) = H_i(C_*)$ for $i < n$ and $H_n(C_*)$ is a quotient of $H_n(C_*^{(n)})$, $b_i = b_i'$ for $i < n$ and $b_n \le b_n'$. $\square$

Let $\chi(M)$ denote the Euler characteristic of M.

(4.2) Corollary $\chi(M) = c_0 - c_1 + c_2 - \cdots.$ $\square$

Adding the inequalities 4.1.n and 4.1.$n - 1$, we obtain:

(4.3) Corollary *For every n, $b_n \le c_n$.* $\square$

The following strengthening of 4.3 is due to E. Pitcher [Pi]. Let t_n equal the number of torsion coefficients in dimension n and let $b_{n-1}(\partial)$ be the rank of the group of boundaries $\partial_n(C_n)$. Then $c_n = b_n + b_n(\partial) + b_{n-1}(\partial)$ and $b_n(\partial) \ge t_n$, *cf.* [ES, V,8.2]. This yields

(4.4) $$c_n \ge b_n + t_n + t_{n-1}.$$

5 Poincaré Duality

As another application we prove the Poincaré duality theorem for a cobordism.

(5.1)   Theorem   *Let $\{V_0, W, V_1)$ be a cobordism. Assume that W is orientable. Then $H_i(W, V_1)$ is isomorphic to $H^{m-i}(W, V_0)$, $i = 0, 1, \ldots m = \dim W$, cohomology with integral coefficients.*

Proof   Consider a presentation $\mathscr{P}$ of the cobordism with handles H_i and the dual presentation $\bar{\mathscr{P}}$ with handles $\bar{H}_j^\lambda$. By 2.3 the handle H_i^λ is the handle $\bar{H}_i^{m-\lambda}$ with the transversal disc becoming the core disc and vice versa. Any choice of orientations for $\mathscr{P}$ and $\bar{\mathscr{P}}$ will produce homology data $C_* = \{C_\lambda, \partial_\lambda\}$, $\bar{C}_* = \{\bar{C}_\lambda, \bar{\partial}_\lambda\}$, where $\partial_\lambda v = \mathfrak{M}_\lambda \cdot v$, $\bar{\partial}_\lambda \bar{v} = \bar{\mathfrak{M}}_\lambda \cdot \bar{v}$. By 3.4 we have $H_*(C_*) = H_*(W, V_0)$ and $H_*(\bar{C}_*) = H_*(W, V_1)$.

Let $C^\lambda = \mathrm{Hom}(C_\lambda, \mathbf{Z})$ and identify C^λ with the free abelian group generated by handles of dimension λ. Then the dual cochain complex $C^* = \{C^\lambda, \delta_\lambda\}$ has the coboundary operator $\delta_\lambda : C^\lambda \to C^{\lambda+1}$,

$$\delta_\lambda v = {}^t\mathfrak{M}_{\lambda+1} \cdot v,$$

and by [D,6.7.11] $H^*(C^*) = H^*(W, V_0)$.

Now, assume that the dual orientation has been chosen for $\bar{\mathscr{P}}$, so that 3.5 holds, and define the homomorphism $g : \bar{C}_* \to C^*$ by $g_\lambda(\bar{H}_i^\lambda) = H_i^{m-\lambda}$. Then the diagram

$$
\begin{array}{ccc}
\bar{C}_\lambda & \xrightarrow{\bar{\partial}_\lambda} & \bar{C}_{\lambda-1} \\
g_\lambda \downarrow & & \downarrow g_{\lambda-1} \\
C^{m-\lambda} & \xrightarrow[\delta_{m-\lambda}]{} & C^{m-\lambda+1}
\end{array}
$$

commutes up to sign: $\delta_{m-\lambda} g_\lambda \bar{H}_i^\lambda = \delta_{m-\lambda} H_i^{m-\lambda} = {}^t\mathfrak{M}_{m-\lambda+1} \cdot H_i^{m-\lambda}$, $g_{\lambda-1} \partial \bar{H}_i^\lambda = g_{\lambda-1}(\bar{\mathfrak{M}} \cdot \bar{H}_i^\lambda) = (-1)^{m-\lambda+1}\, {}^t\mathfrak{M}_{m-\lambda+1} \cdot H_i^{m-\lambda}$.

Thus, g induces an isomorphism $H_\lambda(\bar{C}_*) \simeq H^{m-\lambda}(C^*)$ proving 5.1. $\square$

The Poincaré duality theorem for manifolds with boundary, as in [Sp,6.2.20], is a special case of 5.1 for the cobordism $\{V_0, W, \emptyset\}$. However, our proof applies only to smooth manifolds and the duality isomorphism is defined using a presentation instead of an invariantly defined cap product.

Observe, too, that with W orientable (that is, as everywhere in this book, orientable over the integers), 5.1 remains true for homology and cohomology with coefficients in an arbitrary principal ideal domain R. The only change in the proof is that C^λ is defined as $\mathrm{Hom}_R(C_\lambda, R)$. If W is not orientable, then the theorem remains true with coefficients $\mathbf{Z}/2\mathbf{Z}$. Indeed, the proof in this case is considerably simplified by noticing that with mod 2 intersection numbers Proposition 3.5 becomes trivial.

6 0-Dimensional Handles

The 0-dimensional handles play a somewhat exceptional role in a presentation of a cobordism. For instance, the matrix $\mathfrak{M}_1$ has only zeros and ± 1 as entries. Every row of zeros contributes an infinite cyclic subgroup to $H_0(W, V_0)$, while a row in which there is a nonzero entry contributes nothing. The following theorem shows that it is always possible to find a presentation without superfluous rows.

(6.1) Theorem *Let $\mathscr{C} = \{V_0, W, V_1\}$ be a cobordism. Assume that W is connected. If $V_0 \neq \emptyset$, then there is a presentation of $\mathscr{C}$ without 0-handles. If $V_0 = \emptyset$, then there is a presentation with one 0-handle.*

In both cases, $\mathfrak{M}_1$ is trivial. This is clear if there are no 0-handles; if there is only one 0-handle, then every 1-handle, if any, has both endpoints in the same sphere; thus the relevant intersection number equals 0 and the first level is a handlebody.

Proof Consider first the case $V_0 \neq \emptyset$ and assume that $c_0 > 0$. W_0 is just a disjoint union of $V_0 \times I$ and c_0 copies of D^m. Since $H_0(W_1, V_0) \simeq H_0(W, V_0)$, there must be a 1-handle H^1 with one end in $V^0 \times I$ and another in a 0-handle H^0. Therefore W_1 can be represented as

$$W_1 = (V_0 \times I \cup H^0 \cup H^1) \cup (\text{other 0-handles}) \cup (\text{other 1-handles}).$$

By the Cancellation Lemma, VI,7.4, $V_0 \times I \cup H^0 \cup H^1$ is diffeomorphic to $V_0 \times I$; hence there is a presentation with $c_0 - 1$ 0-handles.

Consider now the case $V_0 = \emptyset$ and assume that $c_0 > 1$. Since W_1 is connected, the same argument as before shows that there is a 1-handle H^1 with ends in different 0-handles H_1^0, H_2^0. Therefore W_1 can be represented as

$$W_1 = (H_1^0 \cup H_2^0 \cup H^1) \cup (\text{other 0-handles}) \cup (\text{other 1-handles}).$$

Again by the Cancellation Lemma, $H_1^0 \cup H_2^0 \cup H^1$ is diffeomorphic to D^m; hence there is a presentation of W_1 with $c_0 - 1$ 0-handles. By induction, there is a presentation of W_1 with one 0-handle. $\square$

(6.2) Corollary *If $V_0 \times I = W_{-1} \subset W_0 \subset W_1 \subset \cdots \subset W_m = W$ is a presentation as in 6.1, then $\pi_i(\partial_+ W_\lambda) \simeq \pi_i(W_\lambda) \simeq \pi_i(W)$ for $i < \lambda < m - i - 1$.*

Proof Applying VI,8.3 successively to the inclusions $W_\lambda \subset W_{\lambda+1} \subset \cdots$ we conclude that

(6.2.1) $\pi_i(W_\lambda) \to \pi_i(W)$ is surjective if $i < \lambda + 1$ and injective if $i < \lambda$.

Similarly, we have from the presentation of the cobordism $\{\partial_+ W_\lambda, V_0\}$ that

(6.2.2) $\quad \pi_i(\partial_+ W_\lambda) \to \pi_i(W_\lambda)$ is surjective if $\lambda < m - i$ and injective if $\lambda < m - i - 1$. $\square$

The case of the fundamental group is of special importance:

(6.3) Corollary *If $\pi_1(W) = 1$, dim $W > 4$, then $\pi_1(\partial_+ W_\lambda) = 1$ for $2 \leq \lambda \leq m - 3$.* $\square$

If W is a compact connected 2-dimensional manifold, then it follows easily from 6.1 that W has a presentation with first level a $(2, 1)$-handlebody W_1 and only one 2-handle. Therefore the boundary of W_1 is a circle, which is a strong restriction. (For instance, if there is only one 1-handle then W must be P^2.) Through a closer inspection of such a presentation one can obtain the classification of compact 2-manifolds.

7 Heegaard Diagrams

Theorem 6.1 has two well known applications. The first one is due to M. Morse [Mo3]:

(7.1) Theorem *On a closed compact connected manifold there is a function with precisely one minimum and one maximum.* $\square$

The other one is due to P. Heegaard:

(7.2) Theorem *Every closed compact connected 3-dimensional manifold M can be obtained by identifying boundaries of two copies B_1, B_2 of the same handlebody B under a diffeomorphism.*

Proof By 6.1 there is a presentation of M with $c_0 = c_3 = 1$ and with the first level a handlebody B_1. Then the first level of the dual presentation is a handlebody B_2 and $M = B_1 \cup_h B_2$, where $h : \partial B_1 \to \partial B_2$ is a diffeomorphism. Since genus and orientability of a (3,1)-dimensional handlebody can be read off from its boundary, B_1 and B_2 are diffeomorphic by VI,11.4. $\square$

Note that by VI,11.4, B is a connected sum along the boundary of g 2-disc bundles over S^1. If M is orientable, then these disc bundles are trivial, i.e., they are solid tori $S^1 \times D^2$.

The minimal genus of B necessary to obtain M is called the *genus of M*. Handlebodies of genus 0 are necessarily homeomorphic to S^3.

Exercise Show that S^3 can be obtained by identifying the boundaries of two handlebodies of arbitrary genus.

In the representation of M given by 7.2 the attaching spheres of 2-handles are precisely the belt spheres (= meridianal circles) of B_2. Thus a system of g disjoint simple closed curves on the boundary of a handlebody B of genus g will determine a manifold M if there is a diffeomorphism of the boundary of B onto itself mapping the system of belt spheres onto the given system of curves. Such a system of curves is called the *Heegaard diagram* of M. Since there is a unique way of attaching a 2-handle along a curve the Heegaard diagram determines M uniquely, at least up to homeomorphism.

Exercise Show that lens spaces $L(p, q)$, as defined in 3.4, are of genus 1 and that the Heegaard diagram of $L(p, q)$ is the (p, q)-torus knot.

Suppose now that M, in addition to the hypotheses of 7.2, is also orientable. Then 7.2 yields a presentation for which matrices $\mathfrak{M}_\lambda$, $\lambda = 1, 3$, have only zero entries, $\mathfrak{M}_2$ is a $g \times g$ matrix, and B is the connected sum along the boundary of g solid tori. Moreover, the elements of $\mathfrak{M}_2$ are intersection numbers of curves in the Heegaard diagram with the meridianal circles of B, and the orientations can be chosen at will. Thus the homology of M is easily computed.

Exercise Find $\mathfrak{M}_3$ if M is non-orientable.

Exercise Show that M is a homology sphere if and only if $\det \mathfrak{M}_2 = \pm 1$.

The fundamental group of M is the same as the fundamental group of the complex consisting of B with 2-discs attached along the attaching spheres of 2-handles (i.e., cores of 2-handles). Therefore it has a presentation with g generators and relations $R_1 = 1, \ldots, R_g = 1$, where R_i is the class of the attaching sphere of the ith 2-handle in $\pi_1(B)$. These can also be easily read from the Heegaard diagram.

Exercise   Show that $\pi_1(L(p, q)) = \mathbf{Z}/p\mathbf{Z}$.

A good illustration of this theory is the following classical construction due to H. Poincaré [P3]. Consider the diagram:

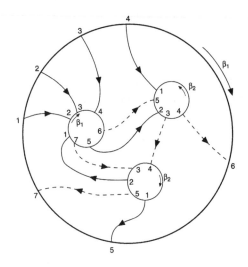

The two circles, both marked β_1, are identified as the arrows indicate as are the circles marked β_2. There results a surface of a handlebody B of genus 2, and the continuous and dotted lines represent simple closed curves α_1, α_2 on it. This is a Heegaard diagram of a manifold M; that is, there is a homeomorphism of ∂B onto itself mapping α_i onto β_i. Poincaré gives two proofs of this: The short and ingenious one consists in noticing that if we cut the surface along α_1, α_2 then the resulting diagram is exactly the same, with the roles of α_1, α_2 and β_1, β_2 interchanged. (A more pedestrian argument would consist in showing that after attaching 2-handles to B along α_1, α_2 the boundary becomes a 2-sphere.)

Exercise Show that M is a homology sphere.

Now, the fundamental group of M has generators g, h and relations

$$g^4 h g^{-1} h = 1, \qquad g^{-1} h g^{-1} h^{-2} = 1.$$

Poincaré shows that it is non-trivial by showing that after adding the relation $g^{-1} h g^{-1} h = 1$ it becomes the icosahedral group. Thus M is not homeomorphic to the 3-sphere.

Poincaré concludes this computation by asking the question: "Is it possible for the fundamental group of M to reduce to the identity element and M not being homeomorphic to the 3-sphere?" After rephrasing this slightly, he ends the paper, his last paper on topology, by saying that "this question would lead us too far."

Indeed, this question, known as the *Poincaré conjecture*, led to a good part of topology created in the 80 years since then. In the next chapter we will present a solution due to S. Smale of a generalization of the Poincaré conjecture in dimensions larger than 4. Recently, M. Freedman [F] solved the 4-dimensional case. But the original question of Poincaré remains unanswered.

8 Historical Remarks

The ideas presented in the last two chapters have a long and somewhat tangled history. Handle presentation of 2-dimensional orientable manifolds appeared for the first time in 1861 in the work of A. Möbius [Mö]. Möbius assumes that there is an imbedding of a closed surface in $\mathbf{R}^3$ such that the height function on it is a Morse function with distinct critical points (in our terminology, of course). This yields a handle decomposition. He then develops a certain notation and an algorithm corresponding essentially to moving handles, which allows him to deduce the analogue of our 7.2 for surfaces. Möbius calls critical points of the height function of index 0 and 2 *elliptic*, those of index 1 *hyperbolic*, and deduces the equality of the Euler characteristic with the alternating sum of numbers of critical points, i.e., 4.2.

Möbius's proofs were grossly deficient. Despite its astonishing novelty his work remained unknown. More details about it can be found in [Pn].

Morse functions reappeared again in Poincaré's "Fifth Complement" [P3]. Poincaré recognized that the topological character of the surface

f = constant does not change between critical points and, in the 3-dimensional case, studied in detail the change at the critical point. The existence of such functions was again assumed.

Poincaré also developed a certain scheme, called by him *the skeleton of a manifold*, which essentially described contiguity relations between handles.

Modern history of handles began with M. Morse's paper [Mo1] in connection with his investigation of critical points of differentiable functions. Since Morse was interested in the homological aspect of the situation, the operation he considered would in today's language be referred to as attaching of cells in the sense of CW-complexes. The differential aspect of the situation seems to have been described first by G. Chogoshvili in [Cho], still in the context of the study of differentiable functions. Thus the Presentation Theorem, 1.2, could have been stated and proved in 1941 by just juxtaposing published papers. But this was not done until 20 years later when various versions of it appeared in the work of S. Smale and A. Wallace. Its real importance became clear when S. Smale used it as a starting point in his successful attack on the Poincaré conjecture in high dimensions [Sm2].

The operation of spherical modifications appeared in differential topology independently of the theory of handles. This technique was introduced independently by A. H. Wallace in [Wa] and J. Milnor in [M7]. (Milnor credits Thom with suggesting the use of the operation.)

There is an inherent difference in the two operations, which may have contributed to their independent appearance. Surgery is informally described as "taking out $S^k \times D^{n+1}$ and gluing in $D^{k+1} \times S^n$." There is not much trouble in endowing the resulting manifold with a smooth structure. On the other hand, when attaching a handle is described as "attaching $D^k \times D^n$ along $S^{k-1} \times D^n$," then the resulting manifolds has "corners" and a device has to be invented to endow it—canonically—with a smooth structure. Complications arise when more than one handle has been attached. When this happens some proofs have to rely strongly on the technique known as vigorous hand waving.

VIII

The h-Cobordism Theorem

The presentation of a cobordism is a geometric object. The homology data derived from it are algebraic objects, subject to algebraic manipulations. In this chapter we study the problem of finding a presentation of a given cobrodism with the minimal number of handles and approach it by trying to realize geometrically certain algebraic operations on incidence matrices. Appropriate conditions for this to be possible are given for elementary row and column operations in Section 1 and, for one further operation corresponding to the cancellation of handles, in Section 2. Section 3 deals with the special case of the matrix $\mathfrak{M}_2$, i.e., of 1-handles. The main result of this chapter, the existence of a minimal presentation for a simply connected cobordism of dimension ≥ 6 with free homology, is proved in Section 4. Among its most important consequences is the Poincaré conjecture for smooth homotopy spheres of dimension at least 5 and the topological characterization of the n-disc, $n \geq 5$, by homotopy conditions. All these results are due to S. Smale.

The relation of h-cobordism is introduced in Section 5. It is an equivalence relation, and equivalence classes of n-dimensional homotopy spheres form a group θ^n. It is shown that for $n \geq 5$ this group is isomorphic to the groups Γ^n and A^n defined in Chapters III and VI respectively. This means that Θ^n

can be identified with the group of differentiable structures on S^n and that each such structure can be realized by an atlas with two charts only.

In Section 6 we obtain a characterization of handlebodies by their homology properties and a description of the structure of highly connected manifolds.

In Section 7 we review some subsequent developments.

1   Elementary Row Operations

Throughout this chapter we make the following assumptions: $\mathscr{C} = \{V_0, W, V_1\}$ is a cobordism with W and V_0 connected. $\mathscr{P}$ is a presentation of $\mathscr{C}$:

$$V_0 \times I = W_{-1} \subset W_0 \subset W_1 \subset \cdots \subset W_m = W, \qquad m = \dim W,$$

with one 0-handle if $V_0 = \emptyset$ and none otherwise (*cf.* VII,6.1), and with homology data $\{C_\lambda, \mathfrak{M}_\lambda\}$, $1 \le \lambda \le m$.

We will consider the following elementary operations on matrices:

E1   Interchange of two columns or rows;

E2   Multiplication of a column or row by -1;

E3   Addition of a column (row) to another column (row).

We say that an operation on a matrix $\mathfrak{M}_\lambda$ yielding $\mathfrak{M}'_\lambda$ can be performed geometrically if there is a presentation $\mathscr{P}'$ of the same cobordism that has $\mathfrak{M}'_\lambda$ as the intersection matrix.

We ask the question: What operations on $\mathfrak{M}_\lambda$ can be performed geometrically? There is certainly no problem with E1: this operation corresponds simply to a renumbering of handles.

To perform E2 note that, whether W is oriented or not, the orientations of cores of handles are always chosen arbitrarily (*cf.* VII,3). But the change of the orientation of the core of the ith λ-handle will change the sign of the ith row of $\mathfrak{M}_{\lambda+1}$ and of the ith column in $\mathfrak{M}_\lambda$. Thus E2 can always be performed geometrically.

We will show presently that E3 can be achieved geometrically. The key to the proof is the following rather obvious lemma.

(1.1)   Lemma   *Let N^{m-1} be a connected closed manifold containing two imbedded $(\lambda - 1)$-spheres S_1, S_2, $1 < \lambda < m$. Assume that S_1 bounds a λ-disc K disjoint from S_2. Then there is an isotopy in N of S_2 to a sphere S, which* can be described as "S_1 joined by a tube to S_2;" that is, S consists of S_1 and S_2 with small discs D_1, D_2 removed and of a tube $\partial D_1 \times I$. If $V^{m-\lambda}$ is a submanifold of N which does not disconnect it, then we can assume that $V \cap S = (V \cap S_1) \cup (V \cap S_2)$.

Proof. Let L be an arc in N with endpoints $\sigma \in S_2$, $s \in S_1$. We request that L be disjoint from V and from the interior of K. Let D_1 be a disc in S_1 centered at s, and let Δ and D_2, $\Delta \subset D_2$, be two concentric discs in S_2 centered at σ. Both D_1 and D_2 should be disjoint from V (see Fig. VIII,1).

The isotopy of S_2 will be performed in two stages.

At the first stage, we move σ along L to s and extend this to an isotopy of M_1 that moves Δ onto D_1, places D_2 in a tubular neighborhood of L, and keeps $S_2 - D_2$ fixed (*cf.* III,3.6).

At the next stage we move D_1 "across K," keeping its boundary fixed, so as to cover $S - \mathring{D}_1$. Again, this is extended to an isotopy of N. Composing these two isotopies results in an isotopy that moves S_2 to a sphere S which consists of $S_2 - D_2$, $S_1 - D_1$, and a tube in a tubular neighborhood of L. $\square$

Let W^m be a manifold, M a connected component of its boundary. In M we have two disjoint oriented $(\lambda - 1)$-spheres Σ_1, Σ_2 and a submanifold $V = V^{m-\lambda}$ which is transverse to both spheres and does not disconnect M. We will assume that the normal bundle to V is oriented. Let $W_2 = W \cup H_1^\lambda \cup H_2^\lambda$, where H_1, H_2 are attached along Σ_1, Σ_2 respectively.

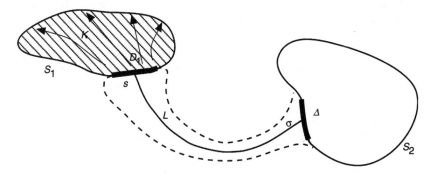

Figure VIII,1

Assuming again $1 < \lambda < m$, we have:

(1.2) Lemma *There is a* $(\lambda - 1)$-*sphere S in M such that:*

  (a) $[S:V] = [\Sigma_2 : V] \pm [\Sigma_1 : V]$;
  (b) W_2 *can be obtained by attaching* H_1^λ *along* Σ_1 *and* H_2^λ *along S.*

Proof We regard W_2 as obtained by two successive attachments of handles: attaching H_1^λ along Σ_1, yielding W_1, followed by attaching H_2^λ along Σ_2. Σ_1 is not present in the boundary N of W_1, but if K is a small disc in N transverse to the belt sphere of H_1^λ, then its boundary S_1 can be viewed as being both in M and in N. Moreover, by VI,6.2, S_1 is "parallel" to Σ_1, i.e., it is a cross section of a tubular neighborhood of Σ_1 in M. Hence S_1 is isotopic to Σ_1 and, with properly chosen orientation, we have

$$[S_1 : V] = [\Sigma_1 : V].$$

As a submanifold of N, S_1 bounds a disc K. Thus we can apply 1.1 to obtain an isotopy (in N) of Σ_2 to a sphere S described there as "Σ_2 joined by a tube to S_1." Since for dimensional reasons S can be assumed to be disjoint from the belt sphere of H_1^λ, it can be regarded as a submanifold of M. Since the intersections of S with V are the same as these of Σ_2 and S_1 we have

$$[S : V] = [\Sigma_2 : V] \pm [S_1 : V],$$

and 1.2(a) follows.

Now, W_2 was obtained by attaching a handle to W_1 along Σ_2. Since Σ_2 is isotopic to S, W_2 can be obtained by attaching a handle along S instead. $\square$

(Observe that V may disconnect M only if $\lambda = 2$. However, 1.2 remains valid even if V disconnects M: the additional intersections will occur in pairs cancelling each other.)

The sign in 1.2 can now be explained by noticing that if $\lambda < m - 1$ then, in the proof of 1.1, in the first stage of the isotopy we can choose whether Δ "moves onto D" with the same or opposite orientation; if $\lambda = m - 1$, then this is predetermined by the chosen orientations. Thus if $\lambda < m - 1$, then the sign in 1.2(a) can be prescribed in advance.

We now collect our results.

(1.3) Theorem *Given* λ, $1 < \lambda < m$, *elementary column operations* $E1$–$E3$ *on* $\mathfrak{M}_\lambda$, *and their inverses, can be performed geometrically affecting only levels* λ *and* $\lambda + 1$.

Proof The operations E1 and E2 having already been dealt with, we will consider E3.

Let $W_{\lambda-1}$ be the $(\lambda - 1)$-st level of the presentation, W_λ the λ-level. Then $W_\lambda = W_{\lambda-1} \cup H_1^\lambda \cup H_2^\lambda \cup \cdots \cup H_{c_\lambda}^\lambda$. Let $V = V^{m-\lambda}$ be the union of all belt spheres of $W_{\lambda-1}$. By VI,9.3, $\partial_+ W_{\lambda-1}$ is connected; we claim that V does not disconnect it. For V might disconnect it only if $\lambda = 2$, but $\partial_+ W_1 - V$ is diffeomorphic to $\partial_+ W_0$ with a finite set of points $(=$ attaching spheres of 1-handles$)$ removed. Since $\partial_+ W_0$ is connected, so also is $\partial_+ W_0 - V$.

This argument shows that 1.2 applies with $M = \partial_+ W_{\lambda-1}$ and $V^{m-\lambda} = V$. It follows that we can obtain W by attaching handles H_1^λ, H_2^λ so that the intersection numbers will be given by 1.2(a), with—if $\lambda < m - 1$—the sign of our choice. Doing this and then attaching the remaining handles will have the effect of adding (or subtracting) the first column of $\mathfrak{M}_\lambda$ to (from) the second. If $\lambda = m - 1$, then we may have to change the orientation of the attaching sphere of H_1^λ beforehand.

This procedure will also affect the matrix $\mathfrak{M}_{\lambda+1}$ (by an appropriate row operation, as it is easy to see), but nothing else. □

(1.4) Corollary *There is a presentation of* $\mathscr{C}$ *in which all matrices* $\mathfrak{M}_\lambda$, $1 < \lambda < m$, *are lower triangular.*

Proof The Euclidean algorithm applied to elements in the first row produces a sequence of column operations yielding one nonzero element in the first row, the greatest common divisor. It is clear how to proceed by induction. □

(1.5) Corollary *If, in addition to our previous assumptions, W is oriented and V_1 is connected, then row operations* $E1$–$E3$ *on matrices* $\mathfrak{M}_\lambda$, $1 < \lambda < m$, *can be performed geometrically.*

Proof 1.3 applies now to the dual presentation $\bar{\mathscr{P}}$ and, by VII,3.5, a column operation on the matrix $\bar{\mathfrak{M}}_{m-\lambda+1}$ of $\bar{\mathscr{P}}$ corresponds to a row operation on $\mathfrak{M}_\lambda$. □

We apply this to obtain a particularly simple presentation of an oriented cobordism, *cf.* [ES, V,8.2].

(1.6) Theorem *If W, V_0, V_1 are connected and oriented, then there is a presentation of W in which all matrices $\mathfrak{M}_\lambda$ are diagonal.*

Proof We proceed by induction. $\mathfrak{M}_1$ is diagonal by VII,6.1; assume that $\mathfrak{M}_i$ is diagonal, $1 \le i \le \lambda < m$. The well-known theorem of H. J. Smith asserts that $\mathfrak{M}_\lambda$ can be reduced to the diagonal form through operations E1–E3 (*cf.* [ST, § 87] or [N,II.15]). We have to see that this can be done without affecting $\mathfrak{M}_{\lambda-1}$. This follows: Since $\{C_\lambda, \mathfrak{M}_\lambda\}$ is a free chain complex, every row of $\mathfrak{M}_\lambda$ containing a nonzero element corresponds to a column of $\mathfrak{M}_{\lambda-1}$ consisting of zeros. Thus we can make all $\mathfrak{M}_i$, $i \le m-1$, diagonal. Finally, $\mathfrak{M}_m$ is diagonal by VII,6.1 again.   □

There is another use of our results. Given a base of a finitely generated free abelian group, any other base can be obtained from the given one by a sequence of the following elementary operations:

   C1 Interchange of two elements;

   C2 Multiplication of an element by -1;

   C3 Addition of an element to another.

Now, if C_λ is the free abelian group generated by λ-handles of a presentation $\mathcal{P}$, then these λ-handles, oriented, constitute a preferred base of C_λ and what we have shown in this section is that any elementary operation C1–C3 on the elements of this base can be realized by an appropriate operation on λ-handles. Thus we have:

(1.7) Proposition *Given a presentation $\mathcal{P}$ of a cobordism with homology data $\{C_\lambda, \mathfrak{M}_\lambda\}$, $1 \le i \le m$, and a base $\mathbf{b}$ of C_λ, there is a presentation $\mathcal{P}'$ that has $\mathbf{b}$ as the preferred base of C_λ.*   □

2 Cancellation of Handles

We will now consider the following operation on matrices:

   E4 Adjoining a row and a column with ± 1 at their intersection and zeros everywhere else.

When performed on a matrix $\mathfrak{M}_\lambda$ of a presentation then one also has to adjoin a row of zeros to $\mathfrak{M}_{\lambda+1}$ and a column of zeros to $\mathfrak{M}_{\lambda-1}$.

It is clear that this operation can be performed geometrically at any level: $W_{\lambda-1} = W_{\lambda-1} \#_b D^m = W_{\lambda-1} \cup H^\lambda \cup H^{\lambda+1}$, by VI,7.4. Moreover the attaching and belt spheres of $H^{\lambda+1}$ and H^λ intersect transversely in one point.

It is quite another matter to perform geometrically the inverse of E4: One has to show that it is possible to remove two handles of successive dimensions knowing only that the algebraic intersection of the corresponding belt and attaching spheres is ±1. Now, it is certainly possible to do so if their geometric intersection consists of only one point; this is precisely the content of VI,7.4. Therefore we are led to the following question: Given two submanifolds V, V' of a manifold M such that $[V:V'] = c$, is it possible to isotope one of them so that it will intersect another in precisely $|c|$ points?

The answer is provided by a theorem of Whitney. We make the following assumptions:

(a) M^{m-1} is a closed connected and simply connected oriented manifold; $V^{\lambda-1}$ and $V^{m-\lambda}$ are closed compact connected submanifolds of M interesecting transversely;

(b) $m \geq 6$; $3 \leq \lambda \leq m - 3$; if $\lambda = 3$, then $\pi_1(M - V^{m-\lambda}) = 1$.

(2.1) Theorem *Under these assumptions there is an isotopy h_t of the identity map of M such that $h_1(V^{\lambda-1})$ intersects $V^{m-\lambda}$ transversely in $\big|[V^{\lambda-1}:V^{m-\lambda}]\big|$ points.*

A proof of this theorem can be found in [M8,6.6]. Its details are subtle but the main idea can be explained rather simply: One chooses a pair of points p, q in the intersection $V^{\lambda-1} \cap V^{m-\lambda}$ with opposite indices and connects them by arcs L_1 in $V^{\lambda-1}$ and L_2 in $V^{m-\lambda}$. The crucial step of the proof consists in imbedding a 2-disc D in M so that its interior is disjoint from $V^{\lambda-1} \cup V^{m-\lambda}$ and its boundary is the simple closed curve $L_1 \cup L_2$. This done, $V^{\lambda-1}$ is moved "across D" so as to remove the pair p, q from the intersection. This isotopy may be assumed to be stationary except in a neighborhood of D. In particular, it moves only a small neighborhood of L_1 and places it in a neighborhood of D.

To apply 2.1 to a cobordism $\mathscr{C} = \{V_0, W, V_1\}$, as in Section 1, we have to make additional assumptions:

(2.2) Proposition *Assume that W is simply connected and $m = \dim W \geq 6$. Given λ, $4 \leq \lambda \leq m - 3$, we can assume that for all i, j the absolute value of $[\alpha_i^{\lambda-1}:\beta_j^{m-\lambda}]$ equals the number of points in $\alpha_i^{\lambda-1} \cap \beta_j^{m-\lambda}$. The same is true if $\lambda = 3$ provided that, in addition, $\pi_1(V_0) = 1$ and there are no 1-handles.*

Proof Let $H^{\lambda-1}$ and H^{λ} be two handles, $\beta^{m-\lambda}$ the belt sphere of $H^{\lambda-1}$, and $\alpha^{\lambda-1}$ the attaching sphere of H^{λ}, $\lambda \geq 4$. Both spheres are in $\partial_+ W_{\lambda-1}$, which is simply connected by VII,6.3. We apply 2.1 to conclude that $\alpha^{\lambda-1}$ can be isotoped to a position in which it will intersect $\beta^{m-\lambda}$ in a number of points equal to the absolute value of $[\alpha^{\lambda-1}:\beta^{m-\lambda}]$. Moreover, this isotopy can be performed in the complement of all other belt and attaching spheres that lie in $\partial_+ W_{\lambda-1}$, this complement being simply connected for dimensional reasons. Thus this operation can be applied to each pair of handles without affecting what has already been achieved.

If $\lambda = 3$ this argument will still work, and 2.1 will apply, for $\partial_+ W_2$ with belt spheres removed is diffeomorphic to $\partial_+ W_1$ with attaching spheres of 2-handles removed. If there are no 1-handles and $\pi_1(V_0) = 1$, then this last manifold is simply connected. $\square$

The last proposition sets the stage for an application of the Cancellation Lemma, VI,7.4, to achieve a simultaneous removal of two handles H_i^{λ} and $H_j^{\lambda-1}$ such that $[\alpha_i^{\lambda-1}:\beta_j^{m-\lambda}] = \pm 1$, which amounts to the inverse of operation E4 on the corresponding incidence matrix. The precise statement follows. W is assumed to be as in 2.2.

(2.3)   Theorem   *Suppose that in the matrix $\mathfrak{M}_{\lambda}$ the ith row and the jth column intersect in ± 1 and have only zero entries elsewhere. If $4 \leq \lambda \leq m - 3$, then one can remove the corresponding λ- and $(\lambda - 1)$-handles from the presentation of W without affecting any other intersection numbers. The same is true if $\lambda = 3$ provided that, in addition, $\pi_1(V_0) = 1$ and there are no 1-handles.*

(Observe that this operation will affect matrices $\mathfrak{M}_{\lambda+1}$, $\mathfrak{M}_{\lambda}$, $\mathfrak{M}_{\lambda-1}$ and that some of them might become trivial. However, any matrix which was trivial before will remain so.)

Proof   Let $i = j = 1$. By 2.2 we can assume that the absolute values of entries in $\mathfrak{M}_{\lambda}$ are actual numbers of points of intersection of respective belt and attaching sphere. In particular, H_1^{λ} is attached away from all handles $H_i^{\lambda-1}$, $i > 1$, and we have

$$W_{\lambda} = W_{\lambda-2} \cup H_1^{\lambda-1} \cup \cdots \cup H_{c_{\lambda-1}}^{\lambda-1} \cup H_1^{\lambda} \cup \cdots \cup H_{c_{\lambda}}^{\lambda}$$

$$= (W_{\lambda-2} \cup H_1^{\lambda-1} \cup H_1^{\lambda}) \cup H_2^{\lambda-1} \cup \cdots \cup H_{c_{\lambda-1}}^{\lambda-1} \cup H_2^{\lambda} \cup \cdots \cup H_{c_{\lambda}}^{\lambda}$$

$$= W_{\lambda-2} \cup H_2^{\lambda-2} \cup \cdots \cup H_{c_{\lambda-1}}^{\lambda-1} \cup H_2^{\lambda} \cup \cdots \cup H_{c_{\lambda}}^{\lambda},$$

the last equality by the Cancellation Lemma, VI, 7.4. $\square$

3 1-Handles

Theorem 2.3 does not apply to $\mathfrak{M}_2$ and in the case of $\mathfrak{M}_3$ it involves an additional unpleasant assumption that there be no 1-handles. In fact, it is not in general possible to eliminate a pair of 1- and 2-handles with the intersection number ± 1. However it is possible to eliminate 1-handles at the expense of replacing them with 3-handles.

(3.1) Proposition *Let $\mathscr{P}$ be a presentation of $\mathscr{C} = \{V_0, W, V_1\}$. Assume that W and V_0 are connected, $\pi_1(W) = 1$, and $\dim W \geq 5$. Then there is a presentation $\mathscr{P}'$ with one 0-handle if $V_0 = \emptyset$, none otherwise, without 1-handles, and with the number of handles of dimensions higher than 3 unchanged.*

Proof By VII,6.1, we can assume that the condition on 0-handles is already satisfied. Note that this implies that $\partial_+ W_0$ is connected. Let H^1 be a 1-handle and L an arc in its boundary intersecting the belt sphere of H^1 transversely in one point. Then the endpoints of L lie in $\partial_+ W_0$. Connecting them there by an arc missing all attaching spheres of other 1-handles (a finite set of points), we obtain a simple closed curve S_1 in $\partial_+ W_1$ intersecting the belt sphere of H^1 transversely in one point and staying away from other 1-handles. We can assume that it is smooth and transverse to all attaching spheres of 2-handles. For dimensional reasons this implies that S_1 is disjoint from all attaching spheres of 2-handles; hence we can view it as being in $\partial_+ W_2$ where, by VII,6.3, it is null-homotopic.

Now, we can represent W_1 as $W_1 \#_b D^m = W_1 \cup H^2 \cup H^3$, where the attaching sphere of H^3 intersects the belt sphere of H^2 transversely in one point and the attaching sphere S_2 of H^2 bounds a 2-disc in $\partial_+ W_1$ that intersects neither the belt spheres of 1-handles nor the attaching spheres of 2-handles. Thus S_1 and S_2 can be regarded as two null-homotopic 1-spheres in $\partial_+ W_2$ and, since $\dim \partial_+ W_2 \geq 4$, they are isotopic there by the theorem of Whitney, II,4.7. Therefore we can assume that H^2 is actually attached along S_1. Symbolically,

$$W_2 = W_0 \cup (\text{1-handles other than } H^1) \cup H^1 \cup H^2 \cup H^3 \cup (\text{2-handles}),$$

where the attaching sphere of H^2 intersects the belt sphere of H^1 transversely in one point. The Cancellation Lemma, VI,7.4, implies now that

$$W_2 = W_0 \cup (\text{1-handles other than } H^1) \cup H^3 \cup (\text{2-handles});$$

i.e., the handle H^1 disappeared from the presentation and a new 3-handle appeared. Repeating this process with other 1-handles will eventually eliminate all of them.   □

The process of elimination of 1-handles will affect intersection matrices. In particular, the matrix $\mathfrak{M}_3$ will become enlarged by addition of new columns corresponding to new 3-handles and, similarly, new rows will appear in $\mathfrak{M}_4$. But since the new 3-handles are attached to $\partial_+ W_1$ and their belts lie in discs, the new rows and columns will have only zero entries.

(3.2) Corollary *If $\mathscr{C}$ is a simply connected 5-dimensional cobordism between two homotopy spheres, then there is a presentation of $\mathscr{C}$ with 2- and 3-dimensional handles only.*

Proof By 3.1 there is a presentation of $\mathscr{C}$ without 0- and 1-handles. Then the dual presentation is without 4- and 5-handles and applying to it 3.1 we obtain what we wanted.   □

This process of trading handles for handles of higher dimension can be applied—under suitable hypotheses—to handles of higher dimension. The essential part of the proof was the construction of the sphere S_1 intersecting the belt of H_1 in one point and null-homotopic in $\partial_+ W_2$; the rest of the argument generalizes immediately.

4 Minimal Presentation; Main Theorems

The main theorem of the theory developed in the last two chapters asserts the existence of a presentation with minimal number of handles. It is due to S. Smale.

Let $\mathscr{C} = \{V_0, W, V_1\}$ be a cobordism. We say that $\mathscr{C}$ is simply connected if V_0, W, V_1 are connected and simply connected. The dimension of $\mathscr{C}$ is by definition the dimension of W. Recall that, for a given presentation, $c_\lambda = \#$ of λ-handles and $b_\lambda(M, V_0) = \text{rank } H_\lambda(M, V_0)$.

(4.1) Theorem *Let $\mathscr{C}$ be a simply connected cobordism of dimension $m \geq 6$ such that $H_i(W, V_0)$ and $H_i(W, V_1)$ are free for $i < k$, $k > 1$. Then there is a presentation of $\mathscr{C}$ such that $c_i = b_i(M, V_0)$ for $i < k$ and $i > m - k$.*

Proof By VII,3.4 we have to show that there is a presentation with the homology data $\{C_i, \mathfrak{M}_i\}$ such that matrices $\mathfrak{M}_i$ are trivial for $i \leq k$ and

$i \geq m - k + 1$. This is done inductively, the cases $i = 1, 2$ being already done in 3.1. Assume then that we already have $\mathfrak{M}_i$ trivial for $i \leq \lambda < k$, $\lambda \geq 2$, and consider $\mathfrak{M}_{\lambda+1}$. By 1.6 we can assume that $\mathfrak{M}_{\lambda+1}$ is diagonal. Since $H_\lambda(W, V_0) = C_\lambda / \mathrm{Im}\, \partial_{\lambda+1}$ is free, the nonzero entries in $\mathfrak{M}_{\lambda+1}$, if any, must equal ± 1. By 2.3, if $\lambda \leq m - 3$ then we can get rid of the corresponding λ-handles without affecting the triviality of $\mathfrak{M}_j$ for $j \leq \lambda$. Therefore there is a presentation with all $\mathfrak{M}_i$, $i \leq \min(k, m - 2)$, trivial.

Consider now the dual presentation. By VII,3.5 this has trivial $\widetilde{\mathfrak{M}}_i$ for $i \geq \max(m - k + 1, 3)$, and that relation is preserved when 1-handles are removed. (As we have noted already, the removal of 1-handles adds only zeros to $\widetilde{\mathfrak{M}}_3$ and $\widetilde{\mathfrak{M}}_4$.) Thus we can assume that $\widetilde{\mathfrak{M}}_1$ and $\widetilde{\mathfrak{M}}_2$ are trivial. Since $H_i(W, V_1)$ is free for $i < k$, we conclude again that it is possible to modify the presentation so that all $\widetilde{\mathfrak{M}}_i$, $i \leq \min(k, m - 2)$, become trivial.   □

The presentation we obtained is, in fact, minimal; this follows from Morse inequalities VII,4.3. Without any assumptions on either $H_i(W, V_0)$ or $H_i(W, V_1)$, the same method of proof leads to the existence of the minimal presentation in the sense of Pitcher inequalities VII,4.4, *cf.* [Sh].

Observe that the hypotheses of Theorem 4.1 are satisfied if $H_i(W)$ is free for $i < k$, $k > 1$, and both V_0 and V_1 are $(k-1)$-connected. This is certainly true if W is closed.

(4.2) Corollary *If W is a simply connected closed manifold of dimension $m \geq 6$ and $H_*(W)$ is free, then there is a handle presentation of W such that $c_\lambda = b_\lambda(W)$, $\lambda = 0, 1, \ldots, m$.*   □

As another corollary of Theorem 4.1 we obtain the main result of this chapter:

(4.3) The h-cobordism Theorem *If $\mathscr{C}$ is a simply connected cobordism of dimension $m \geq 6$ such that $H_*(W, V_0) = 0$, then $\mathscr{C}$ is a trivial cobordism, i.e., W is diffeomorphic to $V_0 \times I$.*

Proof By 4.1 there is a presentation of $\mathscr{C}$ without any handles.   □

The following is known as the Disc Bundle Theorem:

(4.4) Theorem *Let W and ∂W be simply connected and let M be a simply connected closed submanifold in the interior of W. If $\dim M + 3 \leq \dim W \geq 6$*

and $H_*(W, M) = 0$, then W is diffeomorphic to a closed tubular neighborhood T of M in W.

Proof Let $V = \partial T$ and let W_1 be the closure of the complement of T in W. Then $\{V, W_1, \partial W\}$ is a simply connected cobordism, as follows easily from the assumption on the codimension of M and the fact that V is a sphere bundle over M. Evidently, $0 = H_*(W, M) = H_*(W, T) = H_*(W_1, V)$; thus, by 4.3, W_1 is diffeomorphic to $V \times I$. The theorem follows now from VI,5.3. $\square$

Theorem 4.4 yields the following characterization of D^m (take as M a point in the interior of W):

(4.5) Corollary *If W is contractible with a simply connected boundary and of dimension $m \geq 6$, then W is diffeomorphic to D^m.* $\square$

In particular, there is a unique smooth structure on D^m.

The next corollary establishes the Poincaré conjecture for smooth manifolds of dimension larger than 4:

(4.6) Corollary *If M is a homotopy sphere of dimension $m \geq 5$, then M is homeomorphic to S^m.*

Proof By VI,1.4, $M \# (-M)$ bounds a contractible manifold of dimension ≥ 6, thus, by 4.5, diffeomorphic to D^{m+1}. Hence, by VI,2.4, M is homeomorphic to S^m. $\square$

For $m \geq 6$ we can avoid the recourse to VI,2.4: By 4.2 the cobordism $\{\emptyset, M, \emptyset\}$ has a presentation with one 0-handle, one m-handle, and no other handles. Thus $M = D^m \bigcup_h D^m$, where $h: \partial D^m \to \partial D^m$ is a diffeomorphism.

Corollary 4.6 is a case of mixed categories: M is smooth but the conclusion asserts only a topological equivalence. In fact, it is not possible to assert that M is diffeomorphic to S^m: As we will see in X,6, the differential structure on a topological sphere in general is not unique. However, by 4.5, the disc D^m possesses a unique structure.

We have shown that 4.5 in dimension m implies the Poincaré conjecture in dimension $m - 1$. Accordingly, the case $m = 5$ of 4.5 is very difficult. Using a result from Chapter X we can prove the following weaker version.

(4.7) Corollary   *If W is a 5-dimensional contractible manifold bounded by S^4, then W is diffeomorphic to D^5.*

Proof.   Attach to the boundary of W a 5-dimensional disc (i.e., a 5-handle). The resulting manifold W' is a homotopy sphere, which by X,6.3 bounds a contractible manifold. By 4.5, W' is diffeomorphic to S^5, and W—as the complement of the interior of a closed disc in S^5—is diffeomorphic to D^5. $\square$

Exercise   Show that 4.5 in dimension 5 is equivalent to the Poincaré conjecture in dimension 4.

The characterization 4.5 of D^m can be generalized to a characterization of handlebodies.

(4.8) Theorem   *Suppose that M is a simply connected manifold of dimension $m \geq 6$ with a non-empty simply connected boundary. Then M is a handlebody if and only if its reduced homology vanishes in all dimensions but one, and is free.* $\square$

In other words: M is an (m, k)-handlebody of genus g if and only if it has the homology of a wedge of g k-spheres.

Proof.   The condition is necessary by VI,11.5. To prove the sufficiency we consider the cobordism $\{\emptyset, M, \partial M\}$. Since $H_*(M) \simeq H_*(M, \emptyset) \simeq H^*(M, \partial M) \simeq H_*(M, \partial M)$, cf. VII,5.1 and [Sp,V,5.4], $H_*(M, \emptyset)$ and $H_*(M, \partial M)$ are free. Thus 4.1 yields the desired presentation of M. $\square$

Another simple manipulation of duality and universal coefficient theorem yields the following corollary.

(4.9) Corollary   *Suppose that M is a $(k-1)$-connected manifold of dimension $2k \geq 6$ with a non-empty boundary. Then M is a $(2k, k)$-handlebody if and only if the boundary of M is $(k-2)$-connected.* $\square$

A presentation of an (m, k)-handlebody determines a preferred basis for its k-dimensional homology. Conversely, as we have shown in 1.7, every choice of basis can be realized by a presentation.

5 h-Cobordism; The Group θ^m

The notion of h-cobordism which appeared in the name of Theorem 4.3
was introduced by R. Thom in [T5].

Definition A cobordism $\{V_0, W, V_1\}$ is an h-cobordism if the inclusions
$V_0 \hookrightarrow W$, $V_1 \hookrightarrow W$ are homotopy equivalences.

An equivalent requirement is that V_0 and V_1 both the deformation retracts
of W.

An h-cobordism is oriented if W is oriented and $\partial W = V_0 \cup V_1$ as
oriented manifolds. For instance, $\{M \times \{0\}, M \times I, (-M) \times \{1\}\}$ is an oriented cobordism if M is oriented.

If $\{V_0, W, V_1\}$ is an h-cobordism, then each connected component of W
is an h-cobordism between its left and right boundary.

The following lemma explains the apparent asymmetry in the hypotheses
of the h-cobordism theorem, as well as its name:

(5.1) Lemma *A simply connected cobordism* $\{V_0, W, V_1\}$ *is an h-cobordism
if and only if* $H_*(W, V_0) = 0$.

Proof The relative Hurewicz theorem implies that the inclusion $V_0 \hookrightarrow W$
is a weak homotopy equivalence, hence a homotopy equivalence, for W is
a CW-complex (*cf.* [Sp,VII,6.24 and 25]). The conclusion for the other
component follows from VII,5.1.   □

The h-cobordism theorem 4.3 states that simply connected h-cobordisms
of dimension ≥ 6 are trivial, that is, diffeomorphic to products. S. Donaldson
provided examples showing this to be false in dimension 5, *cf.* [DK].
However, according to M. Freedman [F], 5-dimensional h-cobordisms are
homeomorphic to product cobordisms.

Exercise Show that the h-cobordism theorem in dimension 3 is equivalent
to the Poincaré conjecture.

We will now consider compact, closed oriented manifolds of dimension
$m \geq 3$. Two such manifolds V_0, V_1 are said to be h-cobordant if there is
an oriented h-cobordism $\{V_0, W, -V_1\}$. Given two oriented h-cobordisms
$\mathcal{C}$, $\mathcal{C}'$ it easy to see that $\mathcal{C} \cup \mathcal{C}'$ is also an oriented h-cobordism. This shows

that h-cobordism is a transitive relation. It is clearly reflexive and symmetric; thus it is an equivalence relation. More information about the set of its equivalence classes is given by the following three lemmas, in which all manifolds are assumed to be simply connected.

(5.2) Lemma M^m *is h-cobordant to* S^m *if and only if it bounds a contractible manifold.*

Proof Given $\mathscr{C} = \{M^m, W, S^m\}$ and $\mathscr{C}' = \{S^m, D^{m+1}, \emptyset\}$, $\mathscr{C} \cup \mathscr{C}'$ represents M^m as a boundary of a contractible manifold. Conversely, if $M = \partial W$, W contractible, and if W' is obtained from W by removing the interior of an imbedded $(m+1)$-disc, then $\{M, W', S^m\}$ is an h-cobordism. □

It is known [Ma3] that there exist non-simply connected manifolds bounding contractible manifolds. Thus 5.2 does not hold without the assumption of simple connectivity.

(5.3) Lemma *If* M *is h-cobordant to* M_1 *then* $M \# N$ *is h-cobordant to* $M_1 \# N$.

Proof Let $\{M, W, -M_1\}$ and $\{N, N \times I, -N\}$ be oriented h-cobordisms (see Fig. VIII,2). Let L_1 be a neatly imbedded arc in W with endpoints on M and M_1, and let $L = \{\text{pt}\} \times I \subset N \times I$ be an arc in $N \times I$. Recall (*cf.* VI,4) that one can paste W and $N \times I$ along these arcs. The resulting manifold W_1 is simply connected (here we use the assumption $m \geq 3$), one of the components of its boundary is M pasted to N along a point, i.e., it

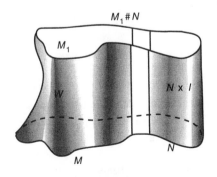

Figure VIII,2

is $M \# N$, and another component is for the same reason $-(M_1 \# N)$. Thus we have a cobordism $\{M \# N, W_1, -(M_1 \# N)\}$. Now, a rather simple computation shows that $H_*(W_1, M \# N) = 0$ and an application of 5.1 concludes the proof.   □

(5.4)   Lemma   *There is a manifold N such that M $\#$ N bounds a contractible manifold if and only if M is a homotopy sphere.*

Proof   If $M \# N$ bounds a contractible manifold, then by 5.2 it is a homotopy sphere. By VI,2.1 both M and N are then homotopy spheres.

Now, if M is a homotopy sphere, then we have already shown in VI,1.4 that $M \# (-M)$ bounds a contractible manifold.   □

Taken together, 5.2–5.4 yield the following:

(5.5)   Theorem   *The set of equivalence classes of the h-cobordism relation is a commutative monoid under the operation of connected sum. The identity element is represented by the class of manifolds bounding a contractible manifold, and the group of invertible elements consists of homotopy spheres.*   □

This group of homotopy spheres will be denoted θ^m. By the h-cobordism theorem and the Poincaré conjecture, 4.6, it can be identified for $m \geq 5$ with the group of smooth structures on the topological m-spheres.

Recall now the group Γ^m of diffeomorphisms of S^{m-1} modulo those which extend over D^m defined in III,6.2 and the group A^m of invertible differentiable structures on a topological m-sphere defined in VI,2. By VI,5.2 there is a monomorphism $\Gamma^m \to A^m$. Assigning to every element of A^m its class in θ^m, we obtain a homomorphism $A^m \to \theta^m$.

(5.6)   Corollary   *For m $\geq$ 5 both homomorphisms $\Gamma^m \to A^m$ and $A^m \to \theta^m$ are isomorphisms.*

It follows that all three groups can be interpreted as groups of differentiable structures on the topological m-sphere, and that every such structure is invertible and can be represented by an atlas with only two charts.

Proof   Consider first the homomorphism $A^m \to \theta^m$. Its kernel consists of those homotopy spheres which bound contractible manifolds; thus, by 4.5, it is injective if $m \geq 5$. Since the homomorphism $\Gamma^m \to A^m$ was shown in VI,5.2 to be injective in all dimensions, all that remains is to show that the composition $\Gamma^m \to A^m \to \theta^m$ is surjective. To see this, let Σ be a homotopy sphere and $D^m \subset \Sigma$ an imbedded m-disc. If D is the complement of its interior, then $\Sigma = D \bigcup_h D^m$, where D is a contractible manifold with boundary S^{m-1}. By 4.5 if $m \geq 6$, and 4.7 if $m = 5$, D is diffeomorphic to D^m, i.e., $\Sigma = \Sigma(h)$. This shows that $\Gamma^m \to \theta^m$ is surjective. $\square$

6   Highly Connected Manifolds

The results of Section 4 can be applied to obtain information about the structure of manifolds that have few nonvanishing homology groups, for instance, highly connected manifolds. More precisely, we will study here connected, closed, orientable manifolds M satisfying the condition:

(k) $H_i(M) = 0$ for $i \neq 0, k, m - k, m$,   where dim $M = m \geq 2k$.

Examples of such manifolds are provided by $(k - 1)$-connected $2k$- and $(2k + 1)$-dimensional manifolds. Another class of examples is obtained by taking two (m, k)-handlebodies with diffeomorphic boundaries and gluing them by means of a diffeomorphism of boundaries. We will show that with some dimensional restrictions, this last example is, in a sense, generic.

First, observe the following:

(6.1)   *If M is orientable and $H_*(M)$ satisfies (k), then all homology groups of M are free except, possibly, $H_k(M)$ if $m = 2k + 1$.*

For if $^t H_i(M)$ stands for the torsion subgroup of $H_i(M)$, then

$$^t H_i(M) = {}^t H^{m-i}(M) = {}^t H_{m-i-1}(M),$$

by VII,5.1 and [Sp,V,5.4].

The following theorem is a generalization of the Heegaard decomposition of 3-dimensional manifolds described in VII,7.2.

(6.2)   Theorem   *Suppose that M is a simply connected, closed manifold satisfying (k). Assume dim $M \geq 5$ and let $b_k = $ rank $H_k(M)$. Then if $m = 2k$, M is a $(2k, k)$-handlebody of genus b_k with a $2k$-disc attached along the boundary. If $m > 2k$, $M = M_1 \bigcup_h M_2$, where M_1 and M_2 are two (m, k)-handlebodies of the same genus and h is a diffeomorphism of their boundaries.*

Proof If $m > 2k + 1$, then we apply 4.2 to obtain the presentation:

$$M = (0\text{-handle}) \cup (b_k \ k\text{-handles})$$

$$\cup \ (b_{m-k} \ (m - k)\text{-handles}) \cup (m\text{-handle}).$$

Since $b_k = b_{m-k}$, the theorem follows in this case by taking as M_1 the kth level of this presentation and as M_2 the kth level of its dual.

If $m = 2k$, then 4.2 yields the presentation:

$$M = (0\text{-handle}) \cup (b_k \ k\text{-handles}) \cup (m\text{-handle}),$$

and the kth level is again the desired handlebody.

If $m = 2k + 1 > 5$, then 4.2 does not apply but 4.1 does and yields a presentation with only k- and $(k + 1)$-handles. If $m = 5$, then this follows from 3.2. Thus, again, we see that $M = M_1 \cup_h M_2$ and, since the boundaries of M_1 and M_2 are homeomomorphic, their genera are equal by VI,11.6.   □

Putting together 6.2 and VI,11.2, we get:

(6.3) Corollary *A $(k - 1)$-connected closed $(2k + 1)$-dimensional manifold, $k > 1$, is obtained by identifying the boundaries of two manifolds, each of which is a connected sum along the boundary of a number of $(k + 1)$-disc bundles over S^k.*   □

If more is known about the manifold, then more can be said about the disc bundles occurring in the decomposition. For instance, if its tangent bundle is stably trivial, then all bundles must be product bundles, as we will see in IX,7.3.

Exercise Show that S^{2k-1} is fibered by $(k - 1)$-spheres over S^k if and only if there exists a $(k - 1)$-connected closed $2k$-manifold M with $H_k(M) = \mathbf{Z}$.

Suppose now that W is a $(2k + 1)$-dimensional $(k - 1)$-connected cobordism between two $(k - 1)$-connected manifolds V_0 and V_1, $k > 1$. Applying to W either 4.1 or 3.2 (if $k = 2$), we obtain a presentation

$$W = (V_0 \times I) \cup (k\text{-handles}) \cup ((k + 1)\text{-handles}) \cup (V_1 \times I),$$

which yields again $W = M_0 \cup_h M_1$ with

$$M_0 = (V_0 \times I) \cup (k\text{-handles}), \qquad M_1 = (V_1 \times I) \cup (k\text{-handles}).$$

Observe now that in both cases the k-handles are attached to $(k-1)$-connected manifolds. Thus VI,11.3 applies and we conclude that M_0 is a boundary connected sum of $V_0 \times I$ and a certain number of $(k+1)$-disc bundles over S^k, all attached to $V_0 \times \{1\}$. The same holds for M_1. We summarize this as follows.

(6.4) Proposition *Under the preceding assumptions, $W = M_0 \cup_h M_1$ where $M_0 = (V_0 \times I) \#_b T_0$, $M_1 = (V_1 \times I) \#_b T_1$, and both T_0 and T_1 are boundary connected sums of a number of disc bundles over S^k.*   □

(6.5) Corollary *Under the preceding assumptions, there are manifolds S_0 and S_1, each a connected sum of a number of k-sphere bundles over S^k, such that $V_0 \# S_0$ and $V_1 \# S_1$ are diffeomorphic.*   □

Again, we will see in the next chapter that if the tangent bundles of W is stably trivial, then S_0 and S_1 are connected sums of product bundles $S^k \times S^k$.

Exercise Show that the boundary of an $(m+1, k)$-handlebody satisfies condition (k).

Exercise Show that if M has a presentation

$M = (0\text{-handle}) \cup (k\text{-handles}) \cup ((k+1)\text{-handles}) \cup ((2k+1)\text{-handle})$

and is a rational homology sphere, then the matrix $\mathfrak{M}_{k+1}$ is square and non-singular. If M is a homotopy sphere, then $\mathfrak{M}_{k+1}$ is unimodular.

7 Remarks

The theorems of this chapter culminate a line of research initiated by Poincaré's question whether vanishing of the fundamental group, a homotopy invariant, characterizes the 3-sphere up to a homeomorphism [P3]. At the time homotopy theory was in its infancy, concurrent with its subsequent development a more general question was asked by W. Hurewicz in his pioneering paper on homotopy groups: Does the homotopy type characterize the homeomorphism type of manifolds?

In this generality the conjecture is already false in dimension 3. For instance, the lens spaces $L(7, 1)$ and $L(7, 2)$ are of the same homotopy type but are not homeomorphic. This follows from the work of Reidemeister,

Moise, and Brody, who classified lens spaces up to homeomorphisms, and J. H. C. Whitehead who gave a classification up to homotopy type.

Simpler—and simply connected—examples were found in higher dimensions, but there still remained the possibility that the homotopy type of the sphere characterizes it up to a homeomorphism. This question, known as the generalized Poincaré conjecture, GPC for short, was finally answered by S. Smale in 1960 for smooth homotopy spheres of dimensions greater than 5 [Sm2]. An authoritative and well-documented account of events surrounding this discovery and a detailed review of results obtained until 1962 can be found in [Sm5,6]. It suffices to say here that A. H. Wallace [Wa] was independently following a similar method and that J. Stallings [St] spurred by the news of Smale's success developed another method, which yielded the proof of GPC for combinatorial homotopy spheres of dimensions larger than 5.

Smale's work culminated in 1962 in his proof of the h-cobordism theorem [Sm3]. Soon afterwards M. Morse undertook to redo Smale's theory. His idea was to try to eliminate directly critical points of the Morse function of a cobordism, instead of eliminating handles. A good account of this is in [B2]. A complete proof of the h-cobordism theorem by this method was given by J. Milnor in 1963 [M8]. Milnor credited the proof of the Cancellation Lemma, in this setting, to M. Morse and qualified it as "quite formidable." (The proof given here in VI,7 does not appear to me "formidable" and might convince the reader of the advantage of the handle approach. The same remark applies to Morse's 30-page proof of VII,7.1 in [Mo3].)

Smale's h-cobordism theorem had three limiting assumptions: simple connectivity, smoothness, and dimension. A counterexample of J. Milnor [M6], which appeared at the same time as the theorem itself, showed that the simple connectivity was essential: $L(7, 1) \times S^4$ and $L(7, 2) \times S^4$ were shown to be h-cobordant but not diffeomorphic. Milnor used the same two lens spaces to provide counterexamples to the so-called *Hauptvermutung* conjecture: Two homeomorphic complexes possess isomorphic subdivisions. (These counterexamples were not manifolds; the question of the validity of the Hauptvermutung for manifolds remained open for some time, see what follows.) An essential role in these constructions was played by Whitehead's torsion invariant. It turned out that this was the right invariant to look at. The first extension of the h-cobordism theorem was given by B. Mazur in [Ma2] in the following form.

A cobordism $\{V_0, W, V_1\}$ is said to be an s-cobordism if the inclusions $V_0 \subset W$, $V_1 \subset W$ are simple homotopy equivalences. The s-cobordism theorem asserts that if dim $W \geq 6$, then an s-cobordism is a product.

The final result combines the ideas of Mazur with those of Stallings and Barden to give a classification of h-cobordisms with the given left-hand boundary V_0. One begins by constructing a group $\text{Wh}(\pi)$ associated to the fundamental group $\pi = \pi_1(V_0)$ and showing that to every h-cobordism $\{V_0, W, V_1\}$ there corresponds an element $\tau(V_0, W)$ of $\text{Wh}(\pi)$, called its torsion. The theorem of Barden–Mazur–Stallings asserts that this correspondence is bijective and trivial cobordisms are those with vanishing torsion. Needless to say, the dimensional restriction of the original h-cobordism theorem must be retained.

The reader of this book should have no trouble in following the presentation of this theory in [K3]. The line of argument is similar to one employed here. The relevant material about simple homotopy theory is in [Co].

The problem of extending the h-cobordism theorem beyond smooth manifolds was very intensely pursued. It appeared quite early that an extension to combinatorial manifolds did not present essential difficulties. Indeed, certain difficulties in Smale's proofs were easy to remedy in the combinatorial setting using older methods and results of J. H. C. Whitehead and M. H. A. Newman. The topological case presented difficulties of another order of magnitude. The first step was due to M. H. A. Newman who in 1966 extended Stallings's engulfing method to topological manifolds, thereby obtaining a proof of GPC for topological manifolds.

A complete theory of topological manifolds was developed by R. Kirby and L. Siebenmann. Besides the topological s-cobordism theorem it includes the theory of surgery and of triangulability. Siebenmann produced a striking example of a topological manifold that does not admit a combinatorial triangulation and an example contradicting the Hauptvermutung for closed manifolds. A summary of these results is in [Si] and a detailed exposition in [KS]. (It is not possible to enumerate in this brief survey many important contributions of a large group of mathematicians. The bibliography in [KS] contains 314 entries.)

In lower dimensions progress was not as rapid. The breakthrough came in 1982 with the work of M. Freedman and S. Donaldson on, respectively, topological and smooth manifolds. We will mention here only those results which are most closely connected to subjects dealt with in this book. By manifold we will mean here a closed, compact, and oriented manifold.

M. Freedman obtained a quite complete topological theory of simply connected 4-dimensional manifolds. The existence and uniqueness theorem of Freedman and Quinn [FQ,10.1] asserts the following:

Suppose we are given a free abelian finitely generated group H, a quadratic form λ on H, and an element ks ∈ Z_2. If λ is even, then we assume that ks = signature λ/8 mod 2.

Then there exists a simply connected 4-dimensional topological manifold M such that $H_2(M) \simeq H$, the intersection form on $H_2(M)$ is isomorphic to λ, and ks = ks(M).

These invariants characterize M up to a homeomorphism.

Here ks(M) stands for the Kirby–Siebenmann obstruction to stable smoothability: ks(M) = 0 if and only if M × **R** is smoothable.

Of course, the Poincaré conjecture for 4-dimensional homotopy spheres is an immediate consequence of the uniqueness part of this theorem. It is also a consequence of the 5-dimensional topological h-cobordism theorem proved by Freedman in [F].

Another consequence is that every simply connected 4-manifold satisfying an additional assumption on its intersection form is a connected sum of a certain number of copies of $S^2 \times S^2$, CP^2 and its conjugate $\overline{CP}^2$, and of two exotic manifolds. (One of these was constructed in VI, 12.)

Freedman also obtained a large number of results without the assumption of simple connectivity. Some of his methods originated in the ideas of A. Casson. For a recent exposition of this theory the reader is referred to [FQ]. An excellent introduction is in [Ki].

The theories of smooth and topological manifolds diverge sharply in dimension 4. In a striking contrast to the results of Freedman, S. Donaldson showed that the intersection form of smooth 4-dimensional manifolds must satisfy strong additional conditions. For instance, if it is definite, then it must be diagonalizable over the integers. Other limitations hold for indefinite forms. They are based on new Donaldson invariants for smooth manifolds. These invariants are of differential-geometric character and originate in theoretical physics.

At this moment there is no existence and uniqueness theorem for smooth 4-manifolds. The situation is quite complicated, as the following two examples of Donaldson show.

First, the 5-dimensional smooth h-cobordism is false: There exist 4-dimensional simply connected smooth manifolds that are h-cobordant but not diffeomorphic. (Note that according to C. T. C. Wall simply connected 4-dimensional manifolds with isomorphic intersection forms are h-cobordant.)

Second, there are non-diffeomorphic smooth structures on $\mathbf{R}^4$. Indeed, Taubes and Gompf constructed uncountable families of such structures.

This theory has interesting ramifications in algebraic geometry. It is expanding rapidly; a recent exposition is in [DK].

In recent years there have been many exciting and profound developments in the theory of 3-dimensional manifolds. In particular, the 3-dimensional Poincaré Conjecture has finally been proved by Grigory Perelman. These developments use methods derived from differential geometry which were originally introduced by Richard Hamilton. For a brief introduction, with bibliography, to this circle of ideas, see John W. Morgan, The Work of Grigory Perelman, Notices of the AMS, Volume **54** (3) pp. 393–399, 2007, reprinted in the present Dover edition as Appendix II on pages 233 through 246.

IX

Framed Manifolds

We will study here a construction due to L. Pontriagin that associates to every map from a manifold M^{k+n} to a sphere S^n a submanifold V^k of M and a framing of its normal bundle.

Section 1 contains some general results concerning the problem of completing a frame field to a framing. In Section 2 we construct the set $\Omega^k(M^{n+k})$ of framed cobordism classes of framed k-dimensional submanifolds of a manifold M^{n+k}, and in Section 3 we show that it can be given a group structure. In Section 4, as an example, we calculate the group $\Omega^0(M^n)$ for M^n a closed, compact, connected, and orientable manifold. The link with homotopy theory is provided in Section 5, where we prove that $\Omega^k(M^{n+k})$ corresponds bijectively to the set $[M^{n+k}, S^n]$ of homotopy classes of maps $M^{n+k} \to S^n$. Using this correspondence in Section 6 we interpret a few standard operations of homotopy theory as operations on framed submanifolds.

In Sections 7 and 8 we study π-manifolds, that is, manifolds that have trivial normal bundle when imbedded in a Euclidean space of high dimension. In Section 7 we establish some critieria for handlebodies to be π-manifolds, and in Section 8 we do the same for some classes of closed manifolds; in particular, we show that homotopy spheres are π-manifolds.

This will be used in the next chapter to establish a link between the groups θ^n and the stable homotopy groups of spheres.

The ideas studied here have an interesting history, which is briefly sketched in Section 9.

1   Framings

An *n-frame* in an *n*-dimensional vector space E is an ordered set of n linearly independent vectors in E, i.e., a basis of E. A frame determines and is determined by a unique isomorphism of E with $\mathbf{R}^n$. Extending these notions to vector bundles, we define a *framing* of an *n*-dimensional vector bundle E to be an ordered set of n everywhere linearly independent sections of E or, equivalently, a continuous map $\mathbf{F}: E \to \mathbf{R}^n$ that is an isomorphism on each fiber. (We could as well define a framing as a section of the associated principal bundle.) We will also have occasion to consider *k-frame fields* in E, such a field being, by definition, an ordered set of k everywhere linearly independent cross sections of E. We assume that the base B of E is a nice space, e.g., a CW-complex.

We will use the same letter to stand for a bundle and its total space. The dimension of a bundle E will mean its fiber dimension; to avoid confusion it will be denoted $\dim_f E$.

A bundle E admits a framing if and only if it is trivial. A framing $\mathbf{F}$ of E determines in every fiber E_p a coordinate system $\mathbf{F}(p)$. This given, every k-frame field $\mathbf{G}$ determines and is determined by a map $h_{\mathbf{F}}(\mathbf{G})$ of the base B of E to the Stiefel manifold $V_{n,k}$ of all k-frames in $\mathbf{R}^n$: The columns of the matrix $h_{\mathbf{F}}(\mathbf{G})(p)$ are the coordinates of vectors of $\mathbf{G}(p)$ in terms of the coordinate system $\mathbf{F}(p)$. We call $h_{\mathbf{F}}(\mathbf{G})$ the *coordinate map* of $\mathbf{G}$ and say that two frame fields are homotopic if their coordinate maps are homotopic. The reader will easily verify that this does not depend of the choice of $\mathbf{F}$ and defines an equivalence relation among k-frame fields. In particular, we have an equivalence relation among the framings of E.

Exercise Show that two framings differing only by an even permutation of cross sections are homotopic.

Exercise Assuming that E is a Riemannian bundle, an orthonormal framing is defined in an obvious way. Show that every framing is homotopic to an orthonormal framing.

If F_1 and F_2 are two frame fields then $F_1 + F_2$ will stand for the ordered set of cross sections of E consisting of the cross sections of F_1 followed by the cross sections of F_2. Of course, $F_1 + F_2$ need not be a frame field.

(1.1) Proposition *Let G be a framing of an n-dimensional bundle E and let G_1 be a k-frame field in E. If* dim $B < n - k$, *then there is an $(n - k)$-frame field G_2 in E such that $G_1 + G_2$ is a framing of E homotopic to G.*

Proof Fix a framing F of E; we will assume all framings and frame fields to be orthonormal. We have the following diagram of maps

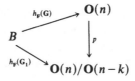

where p is projection of a fibration and $O(n)/O(n - k) = V_{n,k}$ is $(n - k - 1)$-connected [S, 25.6]. Therefore, if dim $B < n - k$, then $p h_F(G)$ is homotopic to $h_F(G_1)$. This homotopy can be covered by a homotopy of $h_F(G)$ to a map h' such that $ph' = h_F(G_1)$. Now, if G' is a framing of E with the coordinate map h' and G_2 the $(n - k)$-field made of the last $n - k$ sections of G', then G' is homotopic to G and $G' = G_1 + G_2$. □

In other words, every framing of a k-dimensional subbundle of E can be completed to a framing of E in a preassigned homotopy class. There is a uniqueness statement for such a completion, which we state in the following form.

(1.2) *Let $E = E_1 + E_2$, where E_1 and E_2 are both trivial bundles, of dimension k and $n - k$ respectively. Suppose that G_1 is a framing of E_1 and G_2, G_2' are two framings of E_2. If* dim $B < n - k - 1$ *and $G_1 + G_2$ is homotopic to $G_1 + G_2'$, then G_2 and G_2' are homotopic (as framings of E_2).*

Proof Let $G = G_1 + G_2$ and consider $h_G(G_1 + G_2)$, a constant map, and $h_G(G_1 + G_2')$. By assumption they are homotopic as maps into $O(n)$. Since $h_G(G_1 + G_2')$ is a composition of $h_G(G_2'): B \to O(n - k)$ with the inclusion $O(n - k) \subset O(n)$, and this inclusion is injective on the homotopy groups in dimensions less than $n - k - 1$, $h_G(G_2')$ is homotopic to a constant as a map to $O(n - k)$. □

(1.3) *Definition* We say that a bundle is *stably trivial* if its Whitney sum with a trivial bundle is trivial. The bundle $TV \oplus \varepsilon^1$ is called the *stable tangent bundle* of V.

Here ε^1 stands for the trivial 1-dimensional vector bundle. Observe that if $V = \partial W$, then $TW \mid V$ can be identified with the stable tangent bundle of V.

(1.4) **Corollary** *If E is a stably trivial bundle over B and dim $B < \dim_f E$, then E is trivial.*

Proof Suppose that $E_1 \oplus E$ is trivial with E_1 trivial; we assume that E_1 and E are Riemannian bundles. Let $\mathbf{G_1}$ be an orthonormal framing of E_1. By 1.1 there is an m-frame field $\mathbf{G}$ in $E_1 \oplus E$, $m = \dim_f E$, such that $\mathbf{G_1} + \mathbf{G}$ is a framing of $E_1 \oplus E$. But then $\mathbf{G}$ is a framing of E. $\square$

It is enough to require that B be of the homotopy type of a CW-complex of dimension satisfying the dimensional restriction. In particular, if either:

(a) B is a connected and compact manifold with non-empty boundary; or

(b) B is the result of removing a point from a compact, connected, and closed manifold;

then it follows from VII,2.2 and VI,8.2 that B is of the homotopy type of a CW-complex of dimension smaller than the dimension of its tangent bundle TB. In this case we can restate 1.4 as follows.

(1.5) **Corollary** *If B satisfies either* (a) *or* (b) *and its tangent bundle is stably trivial, then B is parallelizable.* $\square$

If E is a smooth vector bundle over a smooth manifold, then by a framing of E we will always mean a smooth framing. It follows immediately from I,3.4 that a continuous framing is homotopic to a smooth framing and that all our arguments remain true with all framings, homotopies, and maps assumed smooth.

Consider now a cobordism $\{V_0, W^{k+1}, V_1\}$ where W is assumed to be a submanifold of $\mathbf{R}^{n+k} \times I$ with $V_i \subset \mathbf{R}^{n+k} \times \{i\}$, $i = 0, 1$. By II,3.2 such an imbedding can always be found if $n > k + 1$; henceforth we make this assumption. Let $\mathbf{F_0}$ be a framing of the normal bundle ν of V_0 in $\mathbf{R}^{n+k} \times \{0\}$

and let $\mathbf{E}$ be the standard framing of the tangent bundle of $\mathbf{R}^{n+k+1}$. Since $T\mathbf{R}^{n+k+1}\,|\,V_0 = TW\,|\,V_0 \oplus \nu$, it follows from 1.1 and the assumption $n > k + 1$ that there is a framing $\mathbf{G}_0$ of $TW\,|\,V_0$ such that $\mathbf{F}_0 + \mathbf{G}_0$ is homotopic to $\mathbf{E}\,|\,V_0$.

(1.6) Proposition *If $\mathbf{G}_0$ extends to a framing $\mathbf{G}$ of TW, then $\mathbf{F}_0$ extends to a framing of νW.*

Proof By 1.1 (and the dimensional assumption) there is a framing $\mathbf{F}$ of νW such that $\mathbf{F} + \mathbf{G}$ is homotopic to $\mathbf{E}$. In particular, $(\mathbf{F} + \mathbf{G})\,|\,V_0$ is homotopic to $\mathbf{F}_0 + \mathbf{G}_0$. But $(\mathbf{F} + \mathbf{G})\,|\,V_0 = \mathbf{F}\,|\,V_0 + \mathbf{G}\,|\,V_0 = \mathbf{F}\,|\,V_0 + \mathbf{G}_0$, and it follows from 1.2 that $\mathbf{F}\,|\,V_0$ is homotopic to $\mathbf{F}_0$, provided that $n > k + 1$, as we have assumed. Therefore $\mathbf{F}$ can be viewed as an extension of $\mathbf{F}_0$. $\qquad\square$

2 Framed Submanifolds

We will now concentrate our interest on the framings of the normal bundle of a submanifold.

(2.1) Definition A *framed submanifold* of a manifold M^{n+k} is a pair $(V^k, \mathbf{F})$ where V^k is a neat submanifold of M and $\mathbf{F}$ is a framing of its normal bundle νV^k.

We will allow V^k to be the empty set.

Recall that if $\partial V \neq \emptyset$, then νV restricted to ∂V coincides with the normal bundle of ∂V in ∂M (*cf.* III,4.2). Thus a framing $\mathbf{F}$ of νV induces a framing of the normal bundle of ∂V.

Framed submanifolds arise naturally in the following situation. Suppose that $V = f^{-1}(a)$, where $f: M^{n+k} \to N^n$ is a smooth map and $a \in N^n$ its regular value. Let $p \in V$. As we know (IV,1.4), $Df_p: \nu_p V \to T_a N^n$ is an isomorphism. Therefore if $\mathbf{E}: T_a N^n \to \mathbf{R}^n$ is a system of coordinates—a frame—in $T_a N^n$, then $\mathbf{E} \cdot Df$ is a framing of νV. We will call it the framing induced by Df from $\mathbf{E}$. Note that if two frames $\mathbf{E}_1$ and $\mathbf{E}_2$ yield the same orientation of $T_a N^n$, then the framings induced from them by Df are homotopic.

Among the set of framed submanifolds of M of dimension k we introduce an equivalence relation called *framed cobordism*. First, observe that if $(V, \mathbf{F})$ is a framed submanifold, then $V \times \mathbf{R}$ is a neat submanifold of $M \times \mathbf{R}$ which carries a framing induced by the projection $M \times \mathbf{R} \to M$. The resulting framed manifold will be denoted $(V \times \mathbf{R}, \mathbf{F})$.

Now, let $(V_0, \mathbf{F}_0)$ and $(V_1, \mathbf{F}_1)$ be two framed submanifolds of M. We say that they are *f-cobordant* if there is a framed submanifold $(W^{k+1}, \mathbf{G})$ of $M \times \mathbf{R}$ such that the part of W below the level $t = 0$ coincides with $(V_0 \times \mathbf{R}, \mathbf{F}_0)$ and the part above the level $t = 1$ coincides with $(V_1 \times \mathbf{R}, \mathbf{F}_1)$. Such an f-cobordism is said to be *concentrated between 0 and 1*.

A simple example of f-cobordant framings is provided by homotopic framings of the same submanifold.

(2.2) **Lemma**   *Framed cobordism is an equivalence relation.*

Proof   Only transitivity is non-trivial. So, let $(W_0^{k+1}, \mathbf{G}_0)$ and $(W_1^{k+1}, \mathbf{G}_1)$ be f-cobordisms between $(V_0, \mathbf{F}_0), (V_1, \mathbf{F}_1)$ and $(V_1, \mathbf{F}_1), (V_2, \mathbf{F}_2)$ respectively. Clearly, we can assume that that the cobordism W_0^{k+1} is concentrated between 0 and 1/3 and that W_1^{k+1} is concentrated between 2/3 and 1. But if this is the case, then the manifold $W \subset M \times \mathbf{R}$ defined by

$$W \cap (M \times [-\infty, 2/3]) = W_0^{k+1}, \qquad W \cap (M \times [1/3, \infty]) = W_1^{k+1}$$

realizes the desired f-cobordism between $(V_0, \mathbf{F}_0)$ and $(V_2, \mathbf{F}_2)$. $\square$

If a framed submanifold is moved by an isotopy, then its framing can be pulled along, *cf.* III,2.7.

(2.3) **Proposition**   *If $(V_0, \mathbf{F}_0)$ is a framed submanifold of M and $V_1 \subset M$ is isotopic to V_0, then there is a framing of V_1 such that $(V_0, \mathbf{F}_0)$ is f-cobordant to $(V_1, \mathbf{F}_1)$.*

Proof   Let G be the isotopy, which we view as a level-preserving imbedding $V_0 \times \mathbf{R} \to M \times \mathbf{R}$, and consider the normal bundle ν to $G(V_0 \times \mathbf{R})$ in $M \times \mathbf{R}$. Then $\nu | G(V_0 \times \{i\}) = \nu(V_i), i = 0, 1$,   and $\nu = \pi^*(\nu | G(V_0 \times \{0\}) = \pi^*(\nu(V_0))$, where $\pi: G(V_0 \times \mathbf{R}) \to G(V_0 \times \{0\})$ is the projection. It follows that $(G(V_0 \times \mathbf{R}), \pi^*\mathbf{F}_0)$ is the desired cobordism between $(V_0, \mathbf{F}_0)$ and $(G(V_0 \times \{1\}), \pi^*\mathbf{F}_0 | G(V_0 \times \{1\})) = (V_1, \mathbf{F}_1)$. $\square$

If V^k is a framed submanifold of an oriented manifold M^m, then it is orientable and, in fact, it is possible to orient it unequivocally by agreeing that, at every point, the orientation of V^k followed by the orientation of the frame yields the orientation of M^m; the frame always specifies an orientation. Conversely, we will say that the pair $(V^k, \mathbf{F})$ is oriented if it satisfies the preceding convention. The definition of framed cobordism extends in an obvious way to oriented pairs.

We will show now that every equivalence class of framed k-dimensional submanifolds of $\mathbf{R}^{n+k}$, $n > k + 1$, can be realized by a framing of a connected manifold. This is a consequence of the following:

(2.4) Proposition *Let $(V_0^k, \mathbf{F}_0)$, $(V_1^k, \mathbf{F}_1)$ be two disjoint framed submanifolds of $\mathbf{R}^{n+k}$, $n > k + 1 > 1$. Then there is an imbedding of the connected sum $V_0^k \# V_1^k$ in $\mathbf{R}^{n+k}$ and a framing $\mathbf{F}$ of it which is f-cobordant to the union $(V_0^k, \mathbf{F}_0) \cup (V_1^k, \mathbf{F}_1)$.*

Proof The orientations of V_0 and V_1, necessary to make $V_0 \# V_1$ unambiguous, are chosen so that $(V_0^k, \mathbf{F}_0)$, $(V_1^k, \mathbf{F}_1)$ become oriented pairs. We let $W = (V_0 \times I) \#_b (V_1 \times I)$, the connected sum taken along $V_i \times \{1\}$, $i = 0, 1$. This is unambiguous as well, for the orientations of V_0, V_1 determine the orientations of products with I. By II,3.2 we can view W as a submanifold of $\mathbf{R}^{n+k} \times I$ with $V_0 \cup V_1 \subset \mathbf{R}^{n+k} \times \{0\}$ and $V_0 \# V_1 \subset \mathbf{R}^{n+k} \times \{1\}$. Let $\mathbf{H}_0$ be the framing of $V_0 \cup V_1$ which on V_i equals $\mathbf{F}_i$, $i = 0, 1$. To complete the proof we have to show that $\mathbf{H}_0$ extends over W; this will be done by applying 1.6.

As in 1.6 there is a framing $\mathbf{G}_0$ of $TW|(V_0 \cup V_1)$ such that $\mathbf{H}_0 + \mathbf{G}_0$ is homotopic to $\mathbf{E}$, where $\mathbf{E}$ is the standard framing of $\mathbf{R}^{n+k+1}$. We claim that $\mathbf{G}_0$ extends to a framing of TW. This follows, for $\mathbf{G}_0$ certainly extends to a framing of the tangent bundle of the disjoint union $V_0 \times I \cup V_1 \times I$, which on each oriented chart is homotopic to the framing induced by the chart. Since W is obtained by identification of a chart in $V_0 \times I$ with a chart in $V_1 \times I$ by an orientation preserving diffeomorphism, we can assume that this diffeomorphism identifies the framings as well, i.e., that together they yield a framing of W. $\square$

The last proposition is valid in considerably greater generality: $\mathbf{R}^{n+k}$ can be replaced by an arbitrary connected manifold M^{n+k} and the dimensional restriction $n > k + 1$ reduced to $n > 1$. The proof of this is based on a simple geometric construction but the details are somewhat tedious. We will give a brief sketch.

Consider the $(k + 1)$-dimensional surface S in $\mathbf{R}^{k+1} \times \mathbf{R}^{n-1}$ defined by

$$x_1^2 + \cdots + x_k^2 + 1 = x_{k+1}^2, \qquad x_{k+2} = \cdots = x_{n+k} = 0.$$

It consists of two connected components distinguished by the sign of x_{k+1}. It follows without much trouble from II,2.3 and III,3.7 that there is a chart $U = \mathbf{R}^{k+1} \times \mathbf{R}^{n-1}$ in M^{n+k} that intersects $V_0 \cup V_1$ in S. The required cobordism between $V_0 \cup V_1$ and $V_0 \# V_1$ is now built in $\mathbf{R}^{k+1} \times \mathbf{R}^{n-1} \times I$ in the following way: In $U \times I$ it is the surface T:

$$x_1^2 + \cdots + x_k^2 + 1 = x_{k+1}^2 + 2t,$$

$$x_{k+2} = \cdots = x_{n+k} = 0, \quad t \in I,$$

and in $(M^{n+k} - U) \times I$ it is $(V_0 \cup V_1) \times I$.

Now, this is a rather optimistic description: the two parts do not quite match. They will match if we would take as T the surface

$$\alpha(x_{k+1}, t)(x_1^2 + \cdots + x_k^2) + 1 = x_{k+1}^2 + 2t,$$

$$x_{k+2} = \cdots = x_{n+k} = 0, \quad t \in I,$$

where α is a smooth positive function satisfying:

$$\alpha(x, y) = \begin{cases} 1 & \text{if } |x| < 5, \\ x^2 + 2y - 1 & \text{if } |x| > 10, |y| < 1. \end{cases}$$

T is framed by the normal vector and the standard framing of $\mathbf{R}^{n-1}$. It is not difficult to see that this framing can be assumed to match the obvious framing of $((V_0 \cup V_1) - U) \times I$.

3 $\Omega^k(M^m)$

Let $\Omega^k(M^m)$ be the set of equivalence classes of compact, closed, framed k-dimensional submanifolds of M^m. We will show in Section 5 that $\Omega^k(M^m)$ is in one-to-one correspondence with the set of homotopy classes of maps of M^m to the $(m-k)$-dimensional sphere S^{m-k}. This explains our interest in the possibility of introducing a group structure in $\Omega^k(M^m)$. The obvious way is to try defining addition of two classes as the class of the union of disjoint representatives. Now, by IV,2.4, if $2k < m$, then every two k-dimensional submanifolds can be separated by an isotopy. Thus 2.3 implies that any two classes in $\Omega^k(M^m)$ can be represented by disjoint representatives. A similar argument applied to f-cobordisms shows that if $2k + 1 < m$, then the f-cobordism class of the union of two disjoint framed manifolds depends only on their f-cobordism classes. Therefore, if $2k + 1 < m$, there is a well-defined operation of addition in the set $\Omega^k(M^m)$.

(3.1) Theorem *With this operation of addition $\Omega^k(M^m)$ is an abelian group.*

Proof The operation is certainly associative and commutative and the class of the empty manifold is the neutral element. To show the existence of an inverse, consider first the submanifold $V = V^k \times \{0\}$ of $V^k \times \mathbf{R}^n$ with the framing $\mathbf{E} = (\mathbf{e}_1, \mathbf{e}_2, \ldots, \mathbf{e}_n)$, where $\mathbf{e}_1, \mathbf{e}_2, \ldots, \mathbf{e}_n$ at $p \in V$ is the basis of $\{p\} \times \mathbf{R}^n$ induced by an orthonormal basis of $\mathbf{R}^n$ by the projection $V^k \times \mathbf{R}^n \to \mathbf{R}^n$. We will show that

(i) $(V, \mathbf{E})$ has an inverse.

Let W be the result of imbedding $V \times [0, 1]$ in $V^k \times \mathbf{R}^n \times \mathbf{R}_+$ by the map h,

$$h(p, t) = (p, \cos \pi t, 0, \ldots, 0, \sin \pi t).$$

W is framed by taking as the normal frame $\mathbf{f}_1, \ldots, \mathbf{f}_n$ at $h(p, t)$ the vectors $\mathbf{f}_1 = (\cos \pi t, 0, \ldots, 0, \sin \pi t), \mathbf{f}_2 = \mathbf{e}_2, \ldots, \mathbf{f}_n = \mathbf{e}_n$ (see Fig. IX,1). In particular, at the two components of the boundary of W, $h(V \times \{0\})$ and $h(V \times \{1\})$, this framing becomes, respectively, $\mathbf{e}_1, \mathbf{e}_2, \ldots, \mathbf{e}_n$ and $-\mathbf{e}_1, \mathbf{e}_2, \ldots, \mathbf{e}_n$. Thus ∂W with this framing is null-cobordant, which is the same as saying that $h(V \times \{1\})$ with the framing $-\mathbf{e}_1, \ldots, \mathbf{e}_n$ represents the inverse of $h(V \times \{0\})$ with the framing $\mathbf{e}_1, \ldots, \mathbf{e}_n$. Since $h(V \times \{0\})$ with this framing is clearly f-cobordant to $(V, \mathbf{E})$, this proves (i).

The general case follows now from (i): every framed submanifold $(V^k, \mathbf{F})$ of M^{n+k} may be viewed as the submanifold $(V^k \times \{0\}, \mathbf{E})$ of its tubular neighborhood in M. $\square$

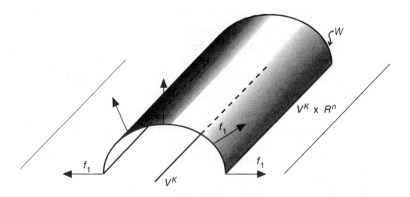

Figure IX,1

Exercise Let $A \in \mathbf{O}(n)$ be a matrix with negative determinant. Show that $(V^k, A \cdot \mathbf{F})$ is the inverse of $(V^k, \mathbf{F})$.

Without the dimensional restriction it is not possible, in general, to introduce a group structure in $\Omega^k(M^m)$: Two framed cobordism classes need not have disjoint representatives and, even if they do, the framed cobordism class of the union may depend on the choice. However, $\Omega^k(S^m)$ carries a group structure with the only restriction being that $m > k + 1$. For in this case we can always isotope representatives of two classes into two different hemispheres and the class of their union will not depend on the isotopies. However, the argument we used to prove the existence of the inverse has to be modified. To do this, we view S^m as a one point compactification of $\mathbf{R}^m$ with $\mathbf{R}^m_+$, $\mathbf{R}^m_-$ corresponding to two hemispheres. Now, let $(V^k, \mathbf{F})$ be a submanifold of $\mathbf{R}^m_+$ and consider its image $(V^k_1, \mathbf{F}_1)$ under the reflection in the last coordinate. A framed cobordism between the two is obtained by imbedding $V^k \times [0, \pi]$ in $\mathbf{R}^{m+1}_+$ by the map

$$(x_1, \ldots, x_{m-1}, x_m, t) \mapsto (x_1, \ldots, x_{m-1}, x_m \cos t, \sin t),$$

$(x_1, \ldots, x_{m-1}, x_m, t) \in V^k \times [0, \pi]$, and suitably framing the result. This shows that $(V^k_1, \mathbf{F}_1)$ is the inverse of $(V^k, \mathbf{F})$.

If we consider oriented pairs, then 3.1 still holds. Indeed, it is clear that there is a bijective correspondence between the set of equivalence classes of oriented pairs and $\Omega^k(M^m)$.

If $\mathbf{F}$ is an oriented framing of an oriented manifold V^k, then the construction employed in the proof of 3.1 yields the inverse of $(V^k, \mathbf{F})$ as a framing of $-V^k$, i.e., V^k with reversed orientation.

Let $\Omega^k_\Sigma(M^m)$, respectively $\Omega^k_S(M^m)$, denote the subset of $\Omega^k(M^m)$ consisting of these classes which can be represented by a framed homotopy sphere, respectively a framed k-sphere. We have by 2.4:

(3.2) Corollary *If $m > 2k + 1 > 2$, then $\Omega^k_\Sigma(S^m)$ and $\Omega^k_S(S^m)$ are subgroups of $\Omega^k(S^m)$.*   □

The generalized version of 2.4 discussed at the end of Section 2 shows that the condition $m > 2k + 1$ can be weakened to $m > k + 1$. Also, with the dimensional restriction retained, S^m can be replaced by an arbitrary connected manifold M^m.

Since S^1 is the only 1-dimensional connected compact closed manifold, we have $\Omega^1_S(M^m) = \Omega^1(M^m)$, $m > 3$.

4 $\Omega^0(M^m)$

There is no method known for an effective calculation of $\Omega^k(M^m)$, $m > 2k + 1$, even if M^m is a sphere. As we will see in the next section this case amounts to a calculation of stable homotopy groups of spheres. Some information can be obtained for low values of k; as an example we will calculate here $\Omega^0(M^m)$ when M^m is closed, compact, connected, and oriented, $m > 1$.

Let $(V, \mathbf{F})$ be a compact framed 0-dimensional submanifold of M. Then V is a finite set of points $p_1, \ldots, p_m$ and the normal bundle of p_i is just the tangent space $T_{p_i}M$. Define ε_i to be $+1$ or -1 according to whether the framing of $T_{p_i}M$ agrees or disagrees with the given orientation of M, and let $\varepsilon(V, \mathbf{F}) = \sum_i \varepsilon_i$; this number will be called the *degree* of $(V, \mathbf{F})$.

We claim that f-cobordant manifolds have the same degree. For let $(W, \mathbf{G})$ be a framed cobordism between $(V, \mathbf{F})$ and $(V', \mathbf{F}')$ concentrated between 0 and 1. Then the part of W between the level $t = 0$ and $t = 1$ consists of a finite number of arcs with endpoints at level 0 or 1. (Here we use the fact that M is closed and compact.) Suppose that both ends p, q of an arc A are at the same level, say $t = 0$. Orient $M \times \mathbf{R}$ by the orientation of M followed by the orientation of $\mathbf{R}$, orient the arc A, and consider the orientation of the tangent space to $M \times I$ along A given by the frame $\mathbf{G} + \tau$, where τ is the oriented tangent vector to A (see Fig. IX,2). This orientation either agrees or disagrees with the orientation of $M \times \mathbf{R}$ along the entire arc A. However, since at the endpoints p, q of A the vectors $\tau(p)$ and $\tau(q)$ point in opposite directions, the framings $\mathbf{G}(p)$ and $\mathbf{G}(q)$ must yield opposite orientations of $M \times \{0\}$, that is, we must have $\varepsilon(p) = -\varepsilon(q)$. A similar argument shows that if p and q are at different levels then $\varepsilon(p) = \varepsilon(q)$. This proves the claim.

Consider now a pair $(p, +1), (q, -1)$. By II,5.3 and 2.3 we can assume that q lies in a tubular neighborhood of p. Then the argument used in the proof of 3.1 shows that this pair is null-cobordant. Therefore if $\varepsilon(V, \mathbf{F}) = 0$, then $(V, \mathbf{F})$ represents the null element of $\Omega^0(M)$.

Taken together, our arguments show that the map $\Omega^0(M) \to \mathbf{Z}$ given by $(V, \mathbf{F}) \mapsto \varepsilon(V, \mathbf{F})$ is a well-defined injective homomorphism. Since dim M is positive, it is also surjective. This we have proved:

(4.1) Proposition *If M is a closed, compact, connected, orientable manifold of dimension > 1, then $\Omega^0(M)$ is isomorphic to the group of integers.* □

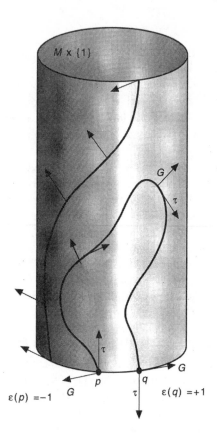

Figure IX,2

Exercise   Show that if M is closed, compact, connected, and non-orientable, then $\Omega^0(M)$ is isomorphic to $\mathbf{Z}_2$.

Some information about $\Omega^1(M^m)$, $m > 3$, can be derived from 3.2. First of all, as we have already noted there, $\Omega^1(M^m) = \Omega_S^1(M^m)$. Second, since dim $M \geq 4$, the isotopy classes of imbeddings $S^1 \to M$ are in one-to-one correspondence with homotopy classes of maps $S^1 \to M$. In particular, if M is simply connected, then there is only one such class and it follows that there is a surjective map $\pi_1(\mathbf{SO}(m-1)) \to \Omega^1(M^m)$. Therefore we have:

(4.2) Proposition   *If M^m is closed, compact, simply connected, and of dimension ≥ 4, then $\Omega^1(M^m)$ has at most two elements.* $\square$

5 The Pontriagin Construction

Let V^k be a neat submanifold of M^{n+k} and N its tubular neighborhood. If the normal bundle of V^k is trivial, then N is diffeomorphic to the product $V^k \times \mathbf{R}^n$; a definite diffeomorphism $t: N \to V^k \times \mathbf{R}^n$ is called a *trivialization* of N. Given a trivialization t, we will define a map $p(V^k, t): M^{n+k} \to S^n$. To do this, we first—and forever—identify $\mathbf{R}^n$ with $S^n - a_-$ via the stereographic projection $p_-: \mathbf{R}^n \to S^n - a_-$, as in I,1.2. This done, $p(V^k, t)$ is defined by

$$p(V^k, t)(q) = \begin{cases} \pi t(q) & \text{if } q \in N, \\ a_- & \text{if } q \notin N, \end{cases}$$

where π is the projection $V^k \times \mathbf{R}^n \to \mathbf{R}^n$.

Since V is assumed to be a neat submanifold of M^{n+k}, the map $p(V^k, t)$ is continuous. Observe that every point x of $\mathbf{R}^n$ is a regular value of $p(V^k, t)$ and that $V_x = p(V^k, t)^{-1}(x)$ is diffeomorphic to V^k. Every point $v \in V_x$ lies on a well-defined fiber N_v of N, which $p(V^k, t)$ maps diffeomorphically onto $\mathbf{R}^n$. Hence there is a unique n-frame in the tangent space of N_v at v that the differential of $p(V^k, t)$ sends to the standard frame of $\mathbf{R}^n$ at x. These frames form a framing $\mathbf{F}_x$ of V_x called the *pull-back framing*. Observe that we defined it not *in abstracto* but as a definite field of frames in TM^{n+k} along V_x.

(5.1) Lemma $(V_x, \mathbf{F}_x)$ *and* $(V_y, \mathbf{F}_y)$ *are f-cobordant.*

Proof Let $h_s(z) = z + s(x - y)$ be an isotopy of $\mathbf{R}^n$. Then x is a regular value of $H = h_s(p(V^k, t))$ and $H^{-1}(x)$ with the pull-back framing is the desired cobordism.   □

Let N and N' be two tubular neighborhoods of V, t and t' their trivializations, and $\mathbf{F}$, $\mathbf{F}'$ the respective pull-back framings of the normal bundle of V. Let $p = p(V, t)$, $p' = p(V, t')$.

(5.2) Lemma *If* $\mathbf{F}$ *is homotopic to* $\mathbf{F}'$, *then* p *is homotopic to* p'.

Proof By the Tubular Neighborhood Theorem, III,3.5, there is an isotopy H_t of the identity map of M^{n+k} such that $H_1 | N: N \to N'$ is linear on each fiber. It follows that $p_1 = p'H_1$ is homotopic to p' and that the pull-back framings $\mathbf{G}$ of $p_1 = p'H_1$ and $\mathbf{F}'$ are homotopic. Let $h_\mathbf{F}(\mathbf{G})$ be the coordinate map as in 1.1. To prove the lemma it is enough to show that if $h_F(G)$ is homotopic to a constant map, then p and p_1 are homotopic.

The maps p and p_1 are defined on the same tubular neighborhood N but using different trivializations: t for p, and $t_1 = t'H_1$ for p_1. The frames of F and G at $v \in V$ can be naturally identified with the two systems of coordinates induced in the fiber of N at v by t and t_1. Therefore we have

$$t_1 = h_F(G)t,$$

where $h_F(G)$ is now interpreted as a map $V \times \mathbf{R}^n \to V \times \mathbf{R}^n$. If F is homotopic to G, then it follows that t_1 is isotopic to t through an isotopy that at each stage is a trivialization of N; hence it can be used to produce a homotopy between p and p_1. $\square$

The following criterion is useful when applying 5.2.

(5.3) *Let F and G be two k-frame fields in $T(V \times \mathbf{R}^n)$ along $V \times \{x\}$. If $F(p) = G(p) \cdot A(p)$, where $A(p)$ is a diagonal matrix with positive entries on the diagonal depending smoothly on $p \in V$ then F and G are homotopic.* $\square$

Now let F be a framing of V^k; F yields a definite diffeomorphism between the total space of the normal bundle of V and $V^k \times \mathbf{R}^n$, which we still denote F. The composition of F with the inverse of an exponential diffeomorphism is then a trivialization t_F of a tubular neighborhood N. The map $p(V^k, t_F): M^{n+k} \to S^n$ is the *Pontriagin map associated to* (V^k, F), and denoted simply $p(V^k, F)$.

There is an ambiguity here in that the trivialization t_F will in general involve, besides the exponential map, some choice of a shrinking map on the normal bundle. However, by 5.2, this will not affect the homotopy class of $p(V^k, F)$: The pull-back framings corresponding to different choices of shrinking maps will also differ only by shrinking, i.e., a multiplication by a diagonal matrix with positive entries, and hence will be homotopic.

Exercise Show that the pull-back framings corresponding to different choices of Riemannian metric are homotopic. (*Hint*: 5.3.)

Observe that $p(V^k, F)^{-1}(0) = V^k$ and that the pull-back framing of V^k corresponding to $p(V^k, F)$ is homotopic to F.

Slightly more general is the following remark: Start with a map $p(V, t)$ and consider the pull-back framing F. Then construct the map $p(V, F)$ and consider its pull-back framing $\mathbf{F}'$. By construction, $\mathbf{F}$ and $\mathbf{F}'$ satisfy 5.3; thus it follows from 5.2 that

(5.4) $p(V, t)$ and $p(V, \mathbf{F})$ are homotopic.

Clearly, if $(W^{k+1}, \mathbf{G})$ is an f-cobordism between $(V_1, \mathbf{F}_1)$ and $(V_2, \mathbf{F}_2)$, then $p(W^{k+1}, \mathbf{G})$ is a homotopy between $p(V_1, \mathbf{F}_1)$ and $p(V_2, \mathbf{F}_2)$. Therefore the Pontriagin construction actually yields a map p of the set $\Omega^k(M^{n+k})$ to the set of homotopy classes of maps $M^{n+k} \to S^n$.

(5.5)   Theorem   *If M^{n+k} is compact and closed, then p is bijective.*

Proof   Let h be a homotopy between $p_0 = p(V_0, \mathbf{F}_0)$ and $p_1 = p(V_1, \mathbf{F}_1)$, and let $z \in \mathbf{R}^n$ be a regular value of h. Then $h^{-1}(z)$ is an f-cobordism between $p_0^{-1}(z)$ and $p_1^{-1}(z)$, all with pull-back framings. By 5.1, $p_0^{-1}(z)$ is f-cobordant to $(V_0, \mathbf{F}_0)$ and $p_1^{-1}(z)$ is f-cobordant to $(V_1, \mathbf{F}_1)$. The three f-cobordisms together show that $(V_0, \mathbf{F}_0)$ is f-cobordant to $(V_1, \mathbf{F}_1)$, i.e., that p is injective.

We will show now that p is surjective, that is, that for every smooth map $f: M^{n+k} \to S^n$ there is a framed submanifold $(V, \mathbf{F})$ such that $p(V, \mathbf{F})$ is homotopic to f.

Let $z \in \mathbf{R}^n$ be a regular value of f. By III,5.1 there is a chart U in the set of regular values and a diffeomorphism $t: f^{-1}(U) \to f^{-1}(z) \times U$ such that the diagram

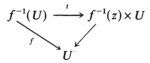

commutes.

Let h_t be a homotopy of the identity map of S^n to a map that maps U diffeomorphically onto $\mathbf{R}^n = S^n - a_-$ and shrinks the complement of U to a_-. Let $g = h_1 f$ and $V = g^{-1}(\mathbf{0})$; since $f^{-1}(U)$ is a tubular neighborhood of V, we have $g = p(V, t)$. Now, g is homotopic to f and, by 5.4, is also homotopic to $p(V, \mathbf{F})$, where $\mathbf{F}$ is the pull-back framing of V corresponding to g. $\square$

As we have noted in 3.1, $\Omega^k(M^{n+k})$ carries a group structure, provided that $n > k + 1$. Thus we have:

(5.6)   Corollary   *If $n > k + 1$ and M^{n+k} is compact and closed, then one can introduce a group structure in the set $[M^{n+k}, S^n]$ of homotopy classes of maps $M^{n+k} \to S^n$.* $\square$

This is the so-called Borsuk–Spanier cohomotopy group of M,

If $M^{n+k} = S^{n+k}$ then, as we know, $\Omega^k(M^{n+k})$ carries a group structure provided only that $n > 1$. A comparison of the definition of addition in this case with the usual definition of addition of maps $S^{n+k} \to S^n$ by concentrating them on different hemispheres shows that $\Omega^k(S^{n+k})$ is simply the homotopy group $\pi_{n+k}(S^n)$.

It follows from 2.4 that the addition in $\Omega^k(M^{n+k})$ can be realized by framing the connected sum:

(5.7) Corollary *Let $p(V_0, \mathbf{F}_0), p(V_1, \mathbf{F}_1): M^{n+k} \to S^n$ be two Pontriagin maps. If $n > k + 1$, then there is a framing $\mathbf{F}$ of $V_0 \# V_1$ such that*

$$p(V_0, \mathbf{F}_0) + p(V_1, \mathbf{F}_1) = p(V_0 \# V_1, \mathbf{F}). \qquad \square$$

Exercise Show that every framed submanifold of $S^m, m > 1$, of codimension 1 is framed null-cobordant. (This accords with the fact that $\pi_m(S^1)$ is trivial for $m > 1$.)

Suppose now that M^n is a closed, compact, connected, and oriented manifold. The orientation determines a generator γ_M of $H_n(M^n)$; similarly, let γ_S be the preferred generator of $H_n(S^n)$ with standard orientation. The degree of a map $f: M^n \to S^n$ was defined in II,2.7 as the integer d such that $f_*\gamma_M = d\gamma_S$. A comparison of II,2.7.1 with the definition of the degree of a framed 0-dimensional submanifold $(V, \mathbf{F})$ of M given in the proof of 4.1 shows that the degree of $(V, \mathbf{F})$ equals the degree of the map $p(V, \mathbf{F})$. Thus we obtain as a corollary of 4.1 and 5.5 the classical theorem of H. Hopf:

(5.8) Corollary *The group $[M^n, S^n]$ is isomorphic to the group of integers, the isomorphism being given by associating to every map its degree.* $\square$

This is equivalent to the statement that $[M^n, S^n]$ is isomorphic to the group of homomorphisms $H_n(M^n) \to H_n(S^n)$. In this form the hypothesis of connectedness is not necessary.

Exercise State and prove the analogue of 5.8 for non-orientable manifolds.

Note that all maps and homotopies in this section can always be taken to be based. For if $p \in M^{n+k}$ and $V^k \subset M^{n+k}$, $p \notin V^k$, then we can always assume that a tubular neighborhood of V^k is disjoint from p and, similarly, if $n > 1$, then every framed cobordism can always be pushed off $\{p\} \times \mathbf{R}$.

6   Operations on Framed Submanifolds and Homotopy Theory

The results of the last section imply that operations on homotopy classes of maps of manifolds into a sphere correspond to operations on framed submanifolds and vice versa. In a number of cases the corresponding operations on framed manifolds have a particularly simple geometric meaning. We will provide a few examples here. For this purpose it will be convenient to consider S^n as the one point compactification of $\mathbf{R}^n$ and, consequently, to study framed submanifolds of $\mathbf{R}^n$.

As usual, $\mathbf{e}_1, \ldots, \mathbf{e}_n$ will stand for the standard framing of the tangent space of $\mathbf{R}^n$ and $\mathbf{n}$ for the inward pointing normal to S^n in $\mathbf{R}^{n+1}$.

(6.1) *Composition of maps.* Consider two maps $g = p(V^m, \mathbf{G})$: $M^{n+k+m} \to S^{n+k}$ and $f = p(V^k, \mathbf{F})$: $S^{n+k} \to S^n$. Then there is a tubular neighborhood $V^m \times \mathbf{R}^{n+k}$ of V^m in M^{n+k+m} on which g is the projection $V^m \times \mathbf{R}^{n+k} \to \mathbf{R}^{n+k}$ and, similarly, we can view f as the projection $V^k \times \mathbf{R}^n \to \mathbf{R}^n$, $V^k \times \mathbf{R}^n \subset \mathbf{R}^{n+k}$. Then $(fg)^{-1}(\mathbf{0}) = g^{-1}(V^k \times \mathbf{0})$ is diffeomorphic to $V^m \times V^k$ and the composition fg is the Pontriagin map associated to a framing of $V^m \times V^k$ imbedded in a tubular neighborhood of V^m. In particular, two such composition maps can be represented by framings of disjoint manifolds, provided that $m < n + k$.

(6.2) *The suspension* Let $f = p(V^k, \mathbf{F})$, where $V^k \subset \mathbf{R}^{n+k}$, i.e., $f: S^{n+k} \to S^n$. Then the suspension of f, Ef, is defined to be the Pontriagin map associated to $(V^k, \mathbf{G})$, where we now view V^k as a submanifold of $\mathbf{R}^{n+k+1}$ and $\mathbf{G} = \mathbf{F} + \mathbf{e}_{n+k+1}$, i.e., $\mathbf{F}$ followed by $\mathbf{e}_{n+k+1}$. It is clear that this defines a homomorphism $E: \pi_{n+k}(S^n) \to \pi_{n+k+1}(S^{n+1})$. The Freudenthal suspension theorem asserts that

(6.2.1)   *E is surjective if $n > k$ and injective if $n > k + 1$.*

Proof Let $p(W^k, \mathbf{G})$ represent an element of $\pi_{n+k+1}(S^{n+1})$, $W^k \subset \mathbf{R}^{n+k+1}$. If $n > k$, then there is an isotopy of $(W^k, \mathbf{G})$ to $(V^k, \mathbf{G}')$, where $V^k \subset \mathbf{R}^{n+k} \subset \mathbf{R}^{n+k+1}$, cf. 2.3 and II,4.7. By 1.1, if $n > k$, then $\mathbf{G}'$ is homotopic to a framing $\mathbf{F} + \mathbf{e}_{n+k+1}$. Thus $p(W^k, \mathbf{G})$ is homotopic to $Ep(V^k, \mathbf{F})$, which proves surjectivity.

To prove that E is injective, assume that $Ep(V^k, \mathbf{F}) = p(V^k, \mathbf{F} + \mathbf{e}_{n+k+1})$ is null-homotopic, i.e., that $(V^k, \mathbf{F} + \mathbf{e}_{n+k+1})$ is a boundary of a framed manifold $(W^{k+1}, \mathbf{G})$, $W^{k+1} \subset \mathbf{R}^{n+k+2}$. A slight extension of the previous argument shows that, without moving V^k, we can isotope W^{k+1} to a position in the subspace defined by $x_{n+k+1} = 0$, $x_{n+k+2} \geq 0$ and that it has there a framing $\mathbf{G}' + \mathbf{e}_{n+k+1}$ such that $\mathbf{G}' | V^k$ is a framing of V^k in $\mathbf{R}^{n+k}$. This shows that $p(V^k, \mathbf{G}' | V^k)$ is null-homotopic. But $\mathbf{G}' | V^k + \mathbf{e}_{n+k+1}$ is homotopic to $\mathbf{F} + \mathbf{e}_{n+k+1}$. Thus by 1.2 $\mathbf{G}' | V^k$ is homotopic to $\mathbf{F}$; hence $p(V^k, \mathbf{F})$ is also null-homotopic. $\square$

6.2.1 shows that the groups $\pi_{n+k}(S^n)$ stabilize for $n > k + 1$: Their value depends only on k and it will be denoted $\pi_k(S)$. By 5.5 this group is isomorphic to $\Omega^k(S^{n+k})$; this will be abbreviated to Ω_f^k and the subgroup $\Omega_S^k(S^{n+k})$ of framed spheres to S_f^k.

(6.3) *The Hopf–Whitehead J-homomorphism* Consider $S^k \subset \mathbf{R}^{k+n}$ with the framing $\mathbf{F}: \mathbf{n}, \mathbf{e}_{k+2}, \ldots, \mathbf{e}_{k+n}$. Let $\gamma: S^k \to \mathbf{SO}(n)$ be a smooth map and $\mathbf{G}$ the framing of S^k whose coordinate map $h_{\mathbf{F}}(\mathbf{G})$ is γ. Then $p(S^k, \mathbf{G})$ maps S^{k+n} to S^n and this construction defines a map $J: \pi_k(\mathbf{SO}(n)) \to \pi_{k+n}(S^n)$. 2.4 implies easily that this map is a homomorphism.

Nothing is gained by considering maps to $\mathbf{O}(n)$ instead of $\mathbf{SO}(n)$, at least if $n > k + 1$. For if $h_{\mathbf{F}}(\mathbf{G})$ has a negative determinant and M is a matrix with determinant -1, then $(M \cdot h_{\mathbf{F}}(\mathbf{G}))^{-1}$ is in $\mathbf{SO}(n)$ and the Pontriagin map associated to it is homotopic to the Pontriagin map associated to $h_{\mathbf{F}}(\mathbf{G})$. (It is the inverse of the inverse if $n > k + 1$.) Hence every framing of $S^k \subset \mathbf{R}^{k+n}$ yields a map in the image of J. Now, if $n > k + 1$, then every imbedding of S^k in $\mathbf{R}^{k+n}$ is isotopic to the standard imbedding. Thus:

(6.3.1) *If $n > k + 1$ and V^k is an imbedded n-sphere in $\mathbf{R}^{k+n}$, then $p(V^k, \mathbf{F})$ is in the image of J, i.e., $S_f^k = \operatorname{Im} J$, cf. 3.2.*

If $k = 1$, then S^1 is the only compact, closed, connected manifold. Thus (*cf.* 4.2):

(6.3.2) *J: $\pi_1(\mathbf{SO}(n)) \to \pi_{n+1}(S^n)$ is surjective for $n > 2$.*

In particular, $\pi_{n+1}(S^n)$ has at most two elements if $n > 2$. (It is somewhat more difficult to prove that it has exactly two elements.) These results appeared already in the first paper of L. Pontriagin on the subject [Po1].

The case $k = 1$, $n = 2$ was first considered by H. Hopf in 1930 [Ho2]. In this case $\mathbf{SO}(2)$ is the group of matrices

$$M(t) = \begin{pmatrix} \cos t & -\sin t \\ \sin t & \cos t \end{pmatrix},$$

and elements of $\pi_1(\mathbf{SO}(2)) = \mathbf{Z}$ can be represented by maps $t \mapsto M(mt)$, $t \in [0, 2\pi]$, $m \in \mathbf{Z}$. The corresponding framing $\mathbf{F}_m$ at the point $(\cos t, \sin t)$ of S^1 is given by the frame

$$v_1 = (\cos mt)\mathbf{n} + (\sin mt)\mathbf{e}_3,$$

$$v_2 = (-\sin mt)\mathbf{n} + (\cos mt)\mathbf{e}_3.$$

It is easy to see that $p(S^1, \mathbf{F}_1)$ is the Hopf map, that is, the projection in the Hopf fibration $S^1 \to S^3 \to S^2$, as given in [S, § 20.1]. It follows from the homotopy exact sequence of this fibration that Hopf map is the generator of $\pi_3(S^2) = \mathbf{Z}$. Thus:

(6.3.3) J: $\pi_1(\mathbf{SO}(2)) \to \pi_3(S^2)$ is an isomorphism.

The maps $p(S^1, \mathbf{F}_m)$ were the first known examples of essential maps $S^{k+n} \to S^n$ with $k > 0$.

Exercise   Consider the cohomotopy group $[M^{n+1}, S^n]$, $n > 2$. Show that if M^{n+1} is 1-connected, then this group has at most two elements, and if M^{n+1} is 2-connected, then it is isomorphic to $\pi_{n+1}(S^n)$. (*Hint*: Consider the maps that factor through S^{n+1}.)

Exercise   Let s: $\pi_k(\mathbf{SO}(n)) \to \pi_k(\mathbf{SO}(n + 1))$ be induced by the inclusion $A \mapsto \left(\begin{smallmatrix} A & 0 \\ 0 & 1 \end{smallmatrix}\right)$. Verify that $EJ = Js$, where E: $\pi_{n+k}(S^n) \to \pi_{n+k+1}(S^{n+1})$ is the suspension homomorphism.

The groups $\pi_k(\mathbf{SO}(n))$ do not depend on n if $n > k + 1$; their common value is denoted $\pi_k(\mathbf{SO})$. The first stable group for a given k is thus $\pi_k(\mathbf{SO}(k + 2))$, which is mapped by the J-homomorphism to $\pi_{2k+2}(S^{k+2})$. The last exercise shows that there is a well-defined stable J-homomorphism J_k: $\pi_k(\mathbf{SO}) \to \pi_k(S)$. Its image is given in 6.3.1; its kernel can be interpreted in the following way.

(6.3.4)   *Let $S^k \subset \mathbf{R}^{n+k}$ be framed by $\mathbf{G}$ and let $h_{\mathbf{E}}\mathbf{G} \in \pi_k(\mathbf{SO}(n))$ be the coordinate map of $\mathbf{G}$ rel. the standard framing $\mathbf{E}$ of S^k in $\mathbf{R}^{n+k}$. Then,*

$h_E\mathbf{G} \in \mathrm{Ker}\, J_k$ *if and only if the framing* $\mathbf{G}$ *extends to a framing of a neat submanifold* $M^{k+1} \subset \mathbf{R}_+^{n+k+1}$ *bounded by* S^k.

It follows that J_k is a monomorphism if and only if every framing of S^k that extends over some manifold in $\mathbf{R}_+^{n+k+1}$, extends over the disc $S_+^{k+1} = \mathbf{R}_+^{n+k+1} \cap S^{k+1}$.

(6.4) *The join* Let $(V^k, \mathbf{F})$ be a framed submanifold of a manifold M^{n+k} and $(W^t, \mathbf{G})$ a framed submanifold of N^{s+t}. Then $V^k \times W^t$ is a submanifold of $M \times N$, with the obvious framing denoted $\mathbf{F} \times \mathbf{G}$. It is easy to see that this is an operation on framed cobordism classes and that it is a bilinear pairing of $[M, S^n] \times [N, S^s]$ to $[M \times N, S^{n+s}]$ when these cohomotopy groups are defined. It can be shown that $Ep(V^k \times W^t, \mathbf{F} \times \mathbf{G})$ is, up to the sign, the so-called join of $p(V^k, \mathbf{F})$ and $p(W^t, \mathbf{G})$ (*cf.* [K1]).

If M and N are Euclidean spaces, then this operation defines a pairing of homotopy groups of spheres, turning the set of stable homotopy groups into a graded ring.

7 π-Manifolds

We have seen that maps $S^{n+k} \to S^n$ can be represented by framed submanifolds of $\mathbf{R}^{n+k}$. We will now focus our attention on the question of which submanifolds of $\mathbf{R}^{n+k}$ can be framed. More precisely, we will study the class of manifolds that can be imbedded in a Euclidean space of a sufficiently high dimension with a trivial normal bundle. Certainly, not all manifolds belong to it. (Already Poincaré noticed that orientability is a necessary condition.) It is somewhat unexpected that this class of manifolds can be characterized by a condition on the tangent bundle.

Recall that a vector bundle is said to be stably trivial if its Whitney sum with a trivial bundle is trivial.

(7.1) *Definition* M is said to be *stably parallelizable* if its tangent bundle is stably trivial.

For brevity we will also call such manifolds π-manifolds. Note that, by 1.4, if $TM \oplus \varepsilon^k$ is trivial, then already $TM \oplus \varepsilon^1$ must be trivial; ε^k denotes as usual the trivial bundle of dimension k. In fact, by 1.5, if M is connected with non-empty boundary, then even TM must be trivial, that is, M must be parallelizable.

Obvious examples of π-manifolds are provided by spheres of all dimensions and by parallelizable manifolds. The latter class includes all Lie groups, in particular $\mathbf{R}^k$ for all k, but among spheres only those of dimension 1, 3, and 7.

It is clear that the boundary of a parallelizable manifold is a π-manifold. The class of π-manifolds behaves nicely under the product operation:

Exercise Show that the product of two π-manifolds is a π-manifold if and only if each factor is a π-manifold.

This property is certainly not shared by parallelizable manifolds: The product of a sphere with a sphere of odd dimension is always parallelizable. For TS^{2k+1} admits a nowhere zero section; thus $TS^{2k+1} = \eta + \varepsilon^1$ for some bundle η, and

$$T(S^{2k+1} \times S^n) = p^*(\eta \oplus \varepsilon^1) \oplus q^*(TS^n) = p^*(\eta) \oplus \varepsilon^1 \oplus q^*(TS^n)$$

$$= p^*(\eta) \oplus q^*(TS^n \oplus \varepsilon^1) = p^*(\eta) \oplus \varepsilon^{n+1}$$

$$= p^*(\eta \oplus \varepsilon^{n+1}) = \varepsilon^{2k+n+1},$$

where p and q are projections.

The normal bundle of a submanifold of a π-manifold has special properties:

(7.2) Theorem (a) *Let N be a π-manifold and $M \subset N$ a submanifold. Then νM is stably trivial if and only if M is a π-manifold.*

(b) *Let N be the total space of a disc bundle over a π-manifold M associated to a vector bundle η. Then N is parallelizable if and only if η is stably trivial.*

Proof (a) Since $TM \oplus \nu M = T_M N,$

$$TM \oplus \nu M \oplus \varepsilon^k = T_M N \oplus \varepsilon^k = \varepsilon^{\dim N+k},$$

and the assertion follows.

(b) We identify M with the zero section of N. Then TN is stably trivial if and only if $TN \mid M$ is stably trivial. Since $TN \mid M = TM \oplus \nu M = TM \oplus \eta$, by III,1.1, and TM is stably trivial, the assertion follows from 1.4. □

In particular, it follows from (a) that M is a π-manifold if and only if it has a trivial normal bundle when imbedded in a Euclidean space of dimension higher than twice the dimension of M. Together with 2.4 this implies that the connected sum of two π-manifolds is a π-manifold.

Comparing 7.2(b), with $M = S^k$, and VI,11.2, we see that an (m, k)-handlebody, $m \geq 2k + 1$, is a π-manifold if and only if it is a boundary connected sum of a number of copies of $D^{m-k} \times S^k$. Appropriate additions can also be made to VIII,6.2–6.3, and VIII,6.4 takes the following form:

(7.3) Proposition *Let $\{V_0, W, V_1\}$ be a $(2k + 1)$-dimensional cobordism. If V_0, W, V_1 are $(k - 1)$-connected and W is a π-manifold, then $W = M_0 \cup_h M_1$, where $M_i = (V_i \times I) \#_b T_i$, $i = 0, 1$, and both T_0 and T_1 are boundary connected sums of a number of copies of $D^{k+1} \times S^k$.*

There is also a criterion for parallelizability of $(2k, k)$-handlebodies. Recall that if M is such a handlebody then a basis of $H_k(M)$ can be represented by imbedded k-spheres, which were called presentation spheres, *cf.* VI,12.

(7.4) Proposition *M is parallelizable if and only if presentation spheres have stably trivial normal bundles.*

Proof The necessity is immediate from 7.2. To prove the sufficiency, let $M = D^{2k} \cup H_1^k \cup \cdots \cup H_s^k$ and let S_i, $i = 1, \ldots, s$, be presentation spheres. We assume that the hemisphere S_{i+} is in D^{2k} and S_{i-} is the core of ith handle H_i^k. If νS_i is stably trivial, which we assume, then $TM \mid S_i$ is stably trivial, hence trivial by 1.4.

Suppose now that we are given a trivialization of $TM \mid D^{2k}$. This induces a trivialization of $TM \mid S_{i+}$. Since $TM \mid S_i$ is trivial, every trivialization of $TM \mid S_{i+}$ extends to a trivialization of the entire bundle $TM \mid S_i$. It follows that $TM \mid (D^{2k} \cup \bigcup_i S_i)$ is trivial. Since $D^{2k} \cup \bigcup_i S_i$ is a deformation retract of M, TM is trivial as well. $\square$

Applying 7.4 to manifolds constructed in VI,12 by the plumbing construction from a graph weighted by elements of $\pi_{k-1}(\mathbf{SO}(k))$, we see that such manifolds are parallelizable if and only if all weights are stably trivial, that is, if they are in the image of ∂: $\pi_k(S^k) \to \pi_{k-1}(\mathbf{SO}(k))$. In particular:

(7.5) *Kervaire manifolds $K(2k)$ and the manifolds $M(4n)$ are parallelizable.*

The Stiefel manifolds $V_{n,k}$ of orthonormal k-frames in $\mathbf{R}^n$ are π-manifolds. This is a consequence of 7.2(b) and the following exercise.

Exercise $V_{n,k+1}$ is a fiber bundle over $V_{n,k}$ with $(n - k - 1)$-spheres as fibers, *cf.* [S, § 7.8]. Show that the associated vector bundle E is stably trivial. (*Hint*: The total space of E can be described as the set of pairs (v, E_v), where $v \in V_{n,k}$ and E_v is the orthogonal complement in $\mathbf{R}^n$ of the subspace spanned by v.)

8 Almost Parallelizable Manifolds

Let M be a $(k - 1)$-connected closed m-dimensional manifold and let M' be M with a disc removed. If $m = 2k + 1$, then 7.3 gives a necessary condition for M' to be parallelizable; if $m = 2k$, then 7.4 gives a necessary and sufficient condition for that. We now ask the question: If M' is parallelizable, is M a π-manifold? This leads to the following definition:

(8.1) *Definition* A bundle ξ over M is *almost trivial* if its restriction to every proper subset of M is trivial. M is *almost parallelizable* if its tangent bundle is almost trivial.

This notion is of interest only if M is compact, connected, and closed. For if M is not connected, then it is almost parallelizable only if it is parallelizable, and the same is true if M is connected with non-empty boundary.

Observe that by 1.4(b) π-manifolds are almost parallelizable. We will study now when is the converse true, that is: When does almost triviality imply stable triviality?

We begin with the general case of an almost trivial bundle ξ of dimension k over a closed orientable manifold M, $m = \dim M$. Assuming then that this is given, let $D \subset M$ be an imbedded m-disc, let M_1 be the closure of $M - D$, and choose framings of $\xi|D$ and $\xi|M_1$. This yields a representation of the total space of ξ as the identification space $D \times \mathbf{R}^k \cup_{\sim} M_1 \times \mathbf{R}^k$, where $(p, v) \in \partial D \times \mathbf{R}^k$ is identified with $(p, \gamma(p) \cdot v) \in \partial M_1 \times \mathbf{R}^k$ for some map $\gamma : \partial D \to \mathbf{SO}(k)$.

(8.2) Lemma *Let $\xi(\gamma)$ be a bundle over S^m with characteristic element $\gamma \in \pi_{m-1}(\mathbf{SO}(k))$. Then $\xi = f^* \xi(\gamma)$ for some map $f : M \to S^m$ of degree* 1.

Proof Identify D with the northern hemisphere D_+ of S^m via a diffeomorphism f. Then the bundle $\xi(\gamma)$ is constructed from the disjoint union $D_+ \times \mathbf{R}^k \cup D_- \times \mathbf{R}^k$ by identifying $(p, v) \in \partial D_+ \times \mathbf{R}^k$ with $(p, \gamma(p) \cdot v) \in \partial D_- \times \mathbf{R}^k$. Now extend f over M_1 by mapping the collar of ∂D in M_1 onto D_- and the rest of M_1 into a single point, the southern pole of S^m. Then f is as desired. $\square$

The lemma implies that all characteristic classes of ξ in dimensions less than m must vanish.

Now, assume that ξ is of dimension m; thus $\gamma \in \pi_{m-1}(\mathbf{SO}(m))$. Since $\xi \oplus \varepsilon^1 = f^*\xi(s_m\gamma)$, ξ is stably trivial if the suspension $s_m\gamma \in \pi_{m-1}(\mathbf{SO}(m+1))$ vanishes. According to Bott, this last group vanishes if $m = 3,5,6,7 \bmod 8$, and is isomorphic to $\mathbf{Z}$ if $m = 4k$, cf. A,5.1. In this case choose a generator η_{4k}. Let $p_k(\xi) \in H^{4k}(M)$ denote the kth Pontriagin class of ξ and let $p_k(\xi)[M] = \langle p_k(\xi), g \rangle$, where g is a generator of $H_{4k}(M)$. It is known, cf. [B3], that:

(8.3) $p_k(\xi(\eta_{4k}))[S^{4k}] = \pm a_k(2k - 1)!$, where $a_k = 2$ if k is odd and 1 if k is even. In particular, the homomorphism $\pi_{m-1}(\mathbf{SO}(m+1)) \to \mathbf{Z}$ given by $\eta \mapsto p_k(\xi(\eta))[S^m]$ is a monomorphism.

We can now collect our results.

(8.4) Proposition *Let ξ be an m-dimensional almost trivial bundle over a manifold of dimension m. If $m = 3,5,6,7 \bmod 8$, then ξ is stably trivial. If $m = 4k$, then ξ is stably trivial if and only if $p_k(\xi) = 0$.*

Proof Only the sufficiency of the condition in the case $m = 4k$ remains to be shown. Assume then that $p_k(\xi) = 0$. By 8.2 $\xi = f^*\xi(\gamma)$. Since f is of degree 1, it follows from the naturality of Pontriagin classes that $p_k(\xi(\gamma)) = 0$; thus $p_k(\xi(s_m\gamma)) = p_k(\xi(\gamma)) = 0$. This implies, by 8.3, that $s_m\gamma = 0$, i.e., $\xi(\gamma)$ is stably trivial. Then ξ is stably trivial itself. $\square$

If ξ is the tangent bundle of a manifold M, we can say more. Let $\sigma(M)$ be the signature of M.

(8.5) Theorem *Let M^m be an almost parallelizable compact closed manifold. If $m \neq 4k$, then M is stably parallelizable. If $m = 4k$, then M is stably parallelizable if and only if $\sigma(M) = 0$.*

Proof Assume that $m = 4k$. By 8.4 we have to show that the signature of M vanishes if and only if $p_k(TM) = 0$. We have already noticed that all classes $p_i(TM)$ with $i < k$ vanish. Therefore, by the Hirzebruch signature theorem [MS,19.4], the signature of M is a nonzero multiple of $p_k(TM)[M]$.

The cases $m = 3,5,6,7 \bmod 8$ were taken care by 8.4; there remain the cases $m = 1,2 \bmod 8$. In these cases, according to a theorem of Adams [Ad], J_{m-1} is a monomorphism; we will dispose of them using 6.3.4.

Let $D \subset M$ be an imbedded m-disc, $S = \partial D$ and $M_1 = M - \text{Int } D$. Imbed M in $\mathbf{R}^{m+n}$, n large, so that $M_1 \subset \mathbf{R}^{m+n}_+$, $D \subset \mathbf{R}^{m+n}_-$, $M \cap \mathbf{R}^{m+n-1} = S$ and $M \pitchfork \mathbf{R}^{m+n-1}$. This is easy to achieve using III,3.7. By 7.2 the normal bundle of M_1 is trivial; let $\mathbf{G}_1$ be a framing of it and $\mathbf{G}$ its restriction to S. By 6.3.4, $\mathbf{G}$ extends over D. Thus the normal bundle of M is trivial and, again by 7.2, M is a π-manifold. $\square$

(8.6) Corollary *Homotopy spheres are π-manifolds.* $\square$

Observe that 8.5 together with 7.4 provides necessary and sufficient conditions for a $2k$-dimensional $(k - 1)$-connected manifold to be stably parallelizable.

Exercise Suppose that $M_1 \# M_2$ is a π-manifold, dim $M_1 \neq 4k$. Show that M_1 and M_2 are both π-manifolds. This is also true if dim $M_1 = 4k$, provided that the signature of M_1 vanishes.

The arguments used in the proof of 8.5 can be refined to yield a theorem of J. Milnor and M. Kervaire [MK], which we will use in X,6.2. Let B_k be the kth Bernoulli number (as in [Hr1,1.5] or [MS]), let j_k be the order of the image of the homomorphism J_k, and let a_k be as in 8.3.

(8.7) Theorem *The signatures of $4k$-dimensional almost parallelizable closed manifolds form a group $t_k\mathbf{Z}$, where*

$$t_k = 2^{2k-1}(2^{2k-1} - 1)B_k j_{4k-1} a_k / k.$$

Proof The signature of the connected sum is the sum of signatures, and the connected sum of almost parallelizable manifolds is almost parallelizable; thus their signatures form a group.

Suppose that M is imbedded in $\mathbf{R}^{m+n}$, n large, $m = \dim M = 4k$, as in the proof of 8.5. Let $\mathbf{G}_1$ be a framing of $M_1 = M \cap \mathbf{R}^{m+n}_+$ and let $\mathbf{G}$ be its restriction to the sphere $S = M \cap \mathbf{R}^{m+n-1}$. By 6.3.4 $h_{\mathrm{E}}(\mathbf{G}) \in \text{Ker } J_{4k-1}$, where $h_{\mathrm{E}}(\mathbf{G}) \in \pi_{4k-1}(\mathbf{SO}(n))$ is the coordinate map of $\mathbf{G}$ rel. the standard framing

E. On the other hand, by 8.2, $\nu = \nu M = f^* \xi(\gamma)$ and a closer look at the proof of 8.2 reveals that $\gamma = h_\mathbf{E}(\mathbf{G})$. Thus γ is a multiple of $j_{4k-1}\eta_{4k}$.

Since f is of degree 1, $p_k(\nu)[M] = p_k(\xi(\gamma))[S^{4k-1}]$, and it follows now from 8.3 that

(*) $p_k(\nu)[M]$ is a multiple of $a_k j_{4k-1}(2k-1)!$.

Now the end is near: By [MS,15.3], $p_k(\nu)[M] = \pm p_k(TM)[M]$ and, by the Signature Theorem,

$$\sigma(M) = p_k(TM)[M]2^{2k}(2^{2k-1}-1)B_k/(2k)!.$$

Together with (*) this shows that $\sigma(M)$ is divisible by t_k.

It remains to be shown that there is an almost parallelizable manifold with signature t_k. To obtain such a manifold, we start by framing the normal bundle of $S^{4k-1} \subset \mathbf{R}^{m+n-1}$ by the generator of Ker J_{4k-1}. This framing extends to a framing of the normal bundle of a manifold $M_1 \subset \mathbf{R}_+^{m+n}$. Attaching a disc to the boundary of M_1 produces a closed manifold M with signature equal to t_k.   □

9 Historical Remarks

The idea that information about a map can be derived from the study of the inverse image of a single point can be traced to the pioneering work of L. E. J. Brouwer, who defined the degree of a map of an n-dimensional manifold to S^n in a way essentially similar to our definition in Section 3, and showed that the degree is an invariant of the homotopy class of the map. Subsequently, H. Hopf showed that it is the only invariant and restated the theory in terms of homology theory [Ho1]. (Both Hopf and Brouwer considered simplicial approximation of continuous maps.) The theorem thus obtained, 5.8 here, was the earliest and the most complete success of homology theory. Only one year later Hopf [Ho2] provided examples showing that the induced homology homomorphism was not sufficient to characterize the homotopy class of a map $S^3 \to S^2$. However, the method he employed could still be called "the method of inverse images": the invariant used to distinguish between non-homotopic maps was the linking number of inverse images of two points. For the map $p(S^1, \mathbf{F}_m)$ from 6.3.3, this is easily seen to equal m.

Hopf's work was very influential and widely known at the time. It is not farfetched to conjecture that it influenced Pontriagin, who announced his idea how to reduce the study of the homotopy groups of spheres to the study of framed submanifolds in a brief note in 1938 [Po1]. Pontriagin hoped that this method would allow the calculation of the stable groups $\pi_{n+k}(S^n)$, n large. He accomplished this for $k = 1$ and (with a mistake) $k = 2$, but the complications in higher dimensions were overwhelming.

The years of war followed and, except for a single paper of B. Eckmann, the method of Pontriagin note of 1938 does not seem to have attracted many followers. Pontriagin published a full description of his method only in 1955 [Po2], one year after the appearance of a paper [T2] of R. Thom who came to the same construction in a different way.

Thom was studying, among other things, the problem of computation of cobordism groups, which he reduced to a computation of homotopy groups of certain spaces. The link between homotopy and cobordism was established by attaching to every homotopy class of maps the inverse image, under a suitably chosen representative, of a fixed submanifold of the target space. "Suitably chosen" means transverse to the submanifold, and in order to prove the existence of such maps Thom had to develop his theory of transversality, another foundational notion of differential topology. This was not necessary for the Pontriagin construction, where the existence of regular values was guaranteed directly by the theorem of Sard and Brown.

In a sense, Thom's method inverted that of Pontriagin, but this time the calculation of the appropriate homotopy groups turned out to be possible. The method was successful and was subsequently applied to the determination of many other types of cobordism. This is presented in [So].

While the study of framed cobordism groups did not provide a method for a calculation of stable homotopy groups of spheres, it did yield some important results when applied to other geometric problems, cf. [K1]. In particular, it allowed an interpretation of the groups of differentiable structures on spheres in terms of stable homotopy groups. The crucial step here was the proof that homotopy spheres are π-manifolds, 8.6 in the preceding. This theory is presented in the next chapter.

J. H. C. Whitehead [Wh2] defined π-manifolds as combinatorial manifolds that have product regular neighborhoods when imbedded in a Euclidean space of sufficiently high dimension. (Regular neighborhood is a combinatorial equivalent of tubular neighborhood.) He observed that if a combinatorial π-manifold is a triangulated smooth manifold then its normal bundle, as defined shortly before by H. Whitney in [Wi1], is trivial.

X

Surgery

The method of surgery was first successfully applied to the investigation of the group θ^n of homotopy spheres [M7, KM2] based on the exact sequence (*cf.* 6.6)

$$0 \to bP^{n+1} \to \theta^n \to \operatorname{Coker} J_n,$$

where J_n is the stable J-homomorphism of IX,6.3, and bP^{n+1} is the group of these homotopy spheres which bound parallelizable manifolds. Now, the image of J_n is a cyclic group of order determined by J. F. Adams; hence θ^n is essentially determined by the stable group $\pi_{n+k}(S^n)$ and by bP^{n+1}. For instance, a proof that bP^{n+1} is finite will show the same for θ^n.

To calculate bP^{n+1} one employs the method of surgery. For suppose that a homotopy sphere Σ is the boundary of a parallelizable manifold M^{n+1} We attempt to find the simplest manifold that Σ bounds, more precisely, to construct a framed cobordism between M and a manifold B still bounded by Σ, but having the simplest possible homology structure. A cobordism is a union of elementary cobordisms and an elementary cobordism is the trace of a surgery (VI,9). Thus we construct the desired cobordism through a sequence of surgeries, each of which aims at eliminating a homology class.

For this to be possible, the homology class has to be represented by an imbedded sphere with a trivial normal bundle, and one has to show that the surgery on this sphere will actually eliminate this homology class without spoiling what has already been achieved.

An appropriate set of conditions for the elimination of a homology class is developed in Section 1. In Section 2 we discuss the problem of framing a surgery. Together with the results of Section 1, this allows us to conclude in 2.2 that a framed surgery on a homology class below the middle dimension is always possible.

The difficulties begin with the surgery on the middle-dimensional homology. In the case of $2k$-dimensional manifolds the difficulty consists in representing the homology class by a sphere with a trivial normal bundle; there is an obstruction to doing this and it provides a monomorphism from bP^{2k} to a finite group. This case is considered in Section 3 for k even and in Section 4 for k odd. One of the consequences is an elucidation in 3.7 and 4.8 of the structure of $(k-1)$-connected $2k$-dimensional π-manifolds.

In the case of odd dimensional manifolds there are no problems with the normal bundle, but special care is necessary to ensure that surgery simplifies homology. We deal with this in Section 5 and prove that it is always possible in this case to find a sequence of surgeries leading to a contractible manifold. Consequently, the group bP^{2k+1} is trivial.

These results are collected in Section 6 and applied to the group θ^n. We prove finiteness in 6.5 and calculate it in a few low dimensional cases. In particular, we obtain examples of nonstandard smooth structures on spheres.

The general line of argument and most of the results of this chapter are due to Milnor and Milnor–Kervaire and come from [M7, KM2], as well as (presumably) from the unpublished second part of [KM2]. However, some of our arguments are quite different from theirs. We use only a very restricted version of the Kervaire invariant and apply in an essential way the theory of handlebodies. This permits a simple treatment of the invariance of the Kervaire invariant in 4.3 and of the odd-dimensional surgery in 5.1. The original proof of 5.1 in [KM2] is a veritable *tour de force* occupying some 16 pages.

Finally, a more detailed account of these and subsequent developments is given in Section 7.

1 Effect of Surgery on Homology

Let M be an m-dimensional manifold and $S \subset M$ a $(k-1)$-dimensional sphere imbedded in the interior of M with a trivial normal bundle. We will now study the effect on M of a surgery on S.

Recall (cf. VI,9) that the trace of a surgery on S is the manifold W obtained from $M \times I$ by attaching a k-dimensional handle to $M \times \{1\}$ along S. Furthermore, $M \times \{0\}$ is referred to as the left-hand boundary $\partial_- W$ and $\chi(M \times \{1\}, S)$ as the right-hand boundary $\partial_+ W$. We will retain this notation even if M has a non-empty boundary, in which case, however, it is no longer true that $\partial W = \partial_+ W \cup \partial_- W$, and consequently W is not a cobordism between $\partial_+ W$ and $\partial_- W$. In order to avoid in the sequel a cumbersome wording caused by this, we will extend the definition of cobordism by saying that M is cobordant to M' if there is a sequence of surgeries leading from M to M'. If M is closed, then this is equivalent to our previous definition by VII,1.1. In any case $\partial M = \partial M'$.

If W is a trace of a surgery, then, by VI,8.1, W is homeomorphic to $(M \times I) \cup_h (D^k \times D^{m-k+1})$, where $h: \partial D^k \times D^{m-k+1} \to M \times \{1\}$ is a diffeomorphism sending $\partial(D^k \times \{0\})$ to S, and the homeomorphism in question is actually a diffeomorphism everywhere except along $\partial D^k \times \partial D^{m-k+1}$. In particular, it makes sense to talk of the transversal disc $D_t = \{0\} \times D^{m-k+1}$ and of the core disc $D_c = D^k \times \{0\}$ as smooth submanifolds of W.

Now, $h(\partial D^k \times D^{m-k+1})$ is a tubular neighborhood of S in M, the sphere $h(p \times \partial D^{m-k+1})$, $p \in \partial D^k$ is called a *meridian*. Any two meridians are isotopic in $M - S$ and, by the Tubular Neighborhood Theorem, the same is true for meridians of two distinct tubular neighborhoods. Note that $h(\partial D_t)$ is isotopic in $\chi(M, S)$ to the meridian $h(p \times \partial D^{m-k+1})$, $p \in \partial D^k$.

The purpose of a surgery on S is to kill the subgroup $[S]$ of $H_{k-1}(M)$ generated by the fundamental class of S. The following proposition gives conditions guaranteeing that this will actually happen.

(1.1) Proposition

$$H_i(\chi(M, S)) = \begin{cases} H_i(M) & \text{if } i < k-1 \text{ and } m \geq 2k-1, \\ H_i(M)/[S] & \text{if } i = k-1 \text{ and } m \geq 2k. \end{cases}$$

Proof Identify M with $M \times \{0\} \subset W$ and $\chi(M, S)$ with $\chi(M \times \{1\}, S) = \partial_+ W$ and consider the diagram

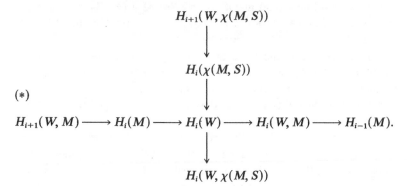

$$H_{i+1}(W, \chi(M, S))$$
$$\downarrow$$
$$H_i(\chi(M, S))$$
$$(\ast) \qquad \qquad \downarrow$$
$$H_{i+1}(W, M) \longrightarrow H_i(M) \longrightarrow H_i(W) \longrightarrow H_i(W, M) \longrightarrow H_{i-1}(M).$$
$$\downarrow$$
$$H_i(W, \chi(M, S))$$

It follows from VI,10.1 that $H_i(W, M) = 0$ unless $i = k$. Thus the inclusion $H_i(M) \to H_i(W)$ is an isomorphism if $i \neq k, k - 1$. Similarly, it follows from VI,9.2 that the inclusion $H_i(\chi(M, S)) \to H_i(W)$ is an isomorphism if $i \neq m - k, m - k + 1$. This yields the first part of the theorem. Now, again from VI,10.1, the image of $H_k(W, M)$ in $H_{k-1}(M)$ under the boundary homomorphism is precisely $[S]$. Since $H_{k-1}(W, M) = 0$, the second part follows. □

In a similar way one can deduce the relation between the homotopy groups of M and $\chi(M, S)$. The case of the fundamental group will be needed later and we state it here.

(1.2) Lemma *If $m > 3$, then $\pi_1(\chi(M, S)) \simeq \pi_1(M)/G$, where G is a normal subgroup of $\pi_1(M)$ containing the homotopy class of S.* □

Proposition 1.1 gives a satisfactory explanation of the effect on the homology of M of surgery below the middle dimension, that is, on a sphere of dimension smaller than $\frac{1}{2}(\dim M - 1)$. To extend our investigation to the case of middle dimension, we study first the case when $m = \dim M = 2k - 1$, in which case the surgery on a $(k - 1)$-dimensional sphere S does not necessarily lead to a disappearance of $[S]$. However, we still have $H_{k-1}(W) \simeq H_{k-1}(M)/[S]$, and since $\chi(M, S) \cup_h D_t$ is a deformation retract of W, the vertical exact sequence in $(\ast)$ yields for $i = k - 1$ the sequence

$$H_k(W, \chi(M, S)) \xrightarrow{\partial} H_{k-1}(\chi(M, S)) \to H_{k-1}(W) \to 0,$$

where the image of ∂ is the homology class generated by $h_*[\partial D_t]$, that is—as we have noted already—the homology class generated by the meridian of S. It follows that if the meridian bounds in $M - S$, then ∂ is trivial and $H_{k-1}(\chi(M, S)) \simeq H_{k-1}(M)/[S]$.

If the meridian bounds in the complement of S, we say that S represents a primitive homology class. A sufficient condition for this is that there be in M an orientable submanifold intersecting S transversely in a single point; this is certainly true if the cohomology class dual to $[S]$ is spherical. Assuming for simplicity that M is closed and letting D stand for the Poincaré duality isomorphism, a necessary and sufficient condition for the homology class α of S to be primitive is that there be a class $\beta \in H_k(M)$ such that $D\alpha \smile D\beta$ generates $H^m(M)$; this happens if and only if α is of infinite order and indivisible.

If dim $M = 2k - 2$, then the vertical exact sequence in (*) becomes, for $i = k - 1$,

$$0 \to H_{k-1}(\chi(M, S)) \to H_{k-1}(W) \to H_{k-1}(W, \chi(M, S)).$$

Since $H_{k-1}(W) \simeq H_{k-1}(M)/[S]$, it follows that $H_{k-1}(\chi(M, S))$ is isomorphic to a subgroup of $H_{k-1}(M)/[S]$, which means that the surgery kills at least the subgroup $[S]$, possibly more. However, there might be trouble in dimension $k - 2$: The vertical exact sequence in (*) becomes, for $i = k - 2$,

$$H_{k-1}(W, \chi(M, S)) \xrightarrow{\partial} H_{k-2}(\chi(M, S)) \to H_{k-2}(W) \to 0,$$

and, since $H_{k-2}(W) \simeq H_{k-2}(M)$, the group $H_{k-2}(\chi(M, S))$ might actually be larger than $H_{k-2}(M)$. We dealt with this problem before: The image of ∂ in $H_{k-2}(\chi(M, S))$ is generated by the meridian of S; therefore if S represents a primitive homology class, ∂ is trivial and $H_{k-2}(\chi(M, S)) \simeq H_{k-2}(M)$.

We now collect our results.

(1.3) Proposition *If* dim $M = 2k - 1$ *or* $2k - 2$ *and S represents a primitive homology class of dimension $k - 1$, then a surgery on S will kill $[S]$ and will not change the homology of M in dimensions less than $k - 1$.* $\square$

Exercise Let M be a $(k - 1)$-sphere bundle over S^k admitting a cross section, and let $S \subset M$ be a fiber. Show that a surgery on S produces a homotopy sphere.

2 Framing of a Surgery; Surgery below Middle Dimension

Assume now that there is given a framing **F** of the stable tangent bundle of M. Let W be the trace of a surgery on M; as before we identify M with $\partial_- W$ and the stable tangent bundles of M and $\chi(M, S)$ with the restrictions of TW to $\partial_- W$ and $\partial_+ W$. Thus **F** becomes a framing of $TW|\partial_- W$ and we will study here the problem of extending it to a framing of TW.

Note that even if the boundary of M is non-empty, the tangent bundle to W is still well-defined, for instance, by viewing $M \times I$ as a subset of $M \times \mathbf{R}$; we will not try to round the corners.

Now, **F** certainly extends over $M \times I$ and to extend it over W it is necessary and sufficient to extend it over the core disc D_c. There is an obstruction to this and it lies in $\pi_{k-1}(\mathbf{SO}(m+1))$. For if **G** is the standard framing of $D^k \times D^{m-k+1}$, then the differential of h sends $\mathbf{G}|\partial D_c$ to a framing of S, and the necessary as well as sufficient condition for **F** to be extendable over D_c is that the map $\omega_h : \partial D_c \to \mathbf{SO}(m+1)$ obtained by comparing $Dh(\mathbf{G}|\partial D_c)$ with $\mathbf{F}|S$ be null-homotopic. (The rows of $\omega_h(p)$ are the coordinates of $Dh(\mathbf{G}|\partial D_c)$ in terms of $\mathbf{F}|S$ at $h(p)$.)

Now, the main property of h is that it yields a surgery on S, i.e., that it maps $S^{k-1} = \partial D_c$ to S. This will still hold if we replace h by $h\gamma$, where $\gamma : \partial D^k \times D^{m-k+1} \to \partial D^k \times D^{m-k+1}$ is given by $\gamma(p, q) = (p, \bar{\gamma}(p) \cdot q)$, $\bar{\gamma} : S^{k-1} \to \mathbf{SO}(m-k+1)$. Note that the differential of γ at $p \in \partial D_c$ is the matrix

$$\begin{pmatrix} I_k & 0 \\ 0 & \bar{\gamma}(p) \end{pmatrix}.$$

Let $\sigma : \mathbf{SO}(m-k+1) \to \mathbf{SO}(m+1)$ be given by

$$A \mapsto \begin{pmatrix} I_k & 0 \\ 0 & A \end{pmatrix}.$$

We then have $\omega_{h\gamma} = \omega_h \cdot \sigma(\bar{\gamma})$ (product of matrices), and, for the homotopy classes,

$$[\omega_{h\gamma}] = [\omega_h] + \sigma_*[\bar{\gamma}], \qquad \sigma_* : \pi_{k-1}(\mathbf{SO}(m-k+1)) \to \pi_{k-1}(\mathbf{SO}(m+1)).$$

Now, if either $m \geq 2k - 1$ or $m = 2k - 2$ but $k \neq 2, 4, 8$, then σ_* is surjective, *cf.* A,5.2, and we can always find γ so that $[\omega_{h\gamma}] = 0$. Thus we have proved the following proposition:

(2.1) Proposition *Given a framing* **F** *of the stable tangent bundle of M and a $(k-1)$-dimensional sphere S in M with a trivial normal bundle, if either*

$m \geq 2k - 1$ *or* $m = 2k - 2$ *but* $k \neq 2, 4, 8$, *then it is possible to perform a surgery on S so that* **F** *will extend to a framing of the tangent bundle of its trace.*   □

Up to now we were considering only tangential framings of W. However, as a result of IX,1.6, analogous results can be obtained for the normal framings. Briefly, if not precisely: If the trace of a surgery on M admits a tangential framing of W, then it admits a normal framing. We will use this in Section 6 to interpret our results in terms of the Pontriagin construction. Until then *framed manifold* will mean manifold with a framing of the stable tangent bundle, and *framed cobordism* will mean cobordism with a framing of the tangent bundle extending the given framing of the stable tangent bundle of the boundary.

We now collect results concerning surgery below the middle dimension.

(2.2)   Theorem   *Let M be a framed m-dimensional π-manifold, $m \geq 2k > 4$. Then there is a framed cobordism between M and a $(k-1)$-connected manifold.*

If $M \neq \emptyset$, then this asserts that there is a sequence of framed surgeries leading to a $(k-1)$-connected manifold. These surgeries are always performed in the interior and affect neither the boundary of M nor its framing. Thus we can view the theorem as asserting that if N bounds a parallelizable manifold M, then N bounds a manifold that is parallelizable and $(k-1)$-connected, $k \leq \frac{1}{2}(\dim N + 1)$.

Proof   We proceed by induction on k. By surgery on 0-spheres, i.e., taking the connected sum of components of M, we can make M connected and we have shown in IX,7 that the connected sum of π-manifolds is a π-manifold. Next is the case of the fundamental group. If $\dim M \geq 3$, then every loop can be represented by a smoothly imbedded 1-dimensional sphere S. By IX,7.2 its normal bundle is stably trivial, hence trivial by IX,1.4. Now 1.2 guarantees that we can eliminate the homotopy class of S by a surgery which, by 2.1, can be framed. Since π_1 is finitely generated, a finite number of surgeries will lead to a simply connected π-manifold.

The inductive step is now clear: If M is $(n-1)$-connected, $n \geq 2$, $m > 2n$, then by the theorem of Hurewicz every homology class in $H_n(M)$ is represented by a map $f: S^n \to M$; by II,3.2 we can assume f to be an imbedding, and the same argument as for the case $n = 1$ shows that $f(S^n)$ has a trivial normal bundle. Therefore we can perform a framed surgery on it and obtain a π-manifold M' such that $H_i(M') = H_i(M) = 0$ for $i < n$ and $H_n(M') = H_n(M)/[S]$. $\square$

This proof illustrates the obstacles we will encounter when trying to perform the surgery in the middle dimensions:

(a) It might not be possible to represent a given homology class by an imbedded sphere with a trivial normal bundle;
(b) Surgery, even if possible, might not achieve the effect of simplifying the homology, cf. 1.3.

The first of these problems will occur in the case of surgery on the middle-dimensional homology of an even-dimensional manifold, the second in the case of odd-dimensional manifolds. Note that in the proof of 2.2 we have shown that if dim $M = 2k + 1$, then every element of $H_k(M)$ can be represented by an imbedded sphere with a trivial normal bundle.

3   Surgery on 4n-Dimensional Manifolds

Let M be an oriented $(k - 1)$-connected $2k$-dimensional manifold, $k \geq 3$. We assume that the boundary of M is either empty or a homotopy sphere; hence $H_k(M)$ is free. By VIII,4.9 and 6.2, either M or M with a disc removed is a $(2k, k)$-handlebody.

By the theorem of Hurewicz, every k-dimensional homology class $x \in H_k(M)$ can be represented by a map $S^k \to M$; by the theorem of Haefliger [H1], this map can be assumed to be an imbedding, which is unique up to an isotopy if $k \geq 4$. Let $S(x)$ be the (oriented) k-sphere in M representing x. We assume that its normal bundle is oriented by the convention adopted in IV,5. Hence the function $x \mapsto \alpha(x)$ that assigns to x the characteristic element $\alpha(x) \in \pi_{k-1}(\mathbf{SO}(k))$ of the normal bundle of $S(x)$ is well-defined. Since $\pi_2(\mathbf{SO}(3)) = 0$, it is defined for $k = 3$ as well.

Let $x, y \in H_k(M)$ be represented by imbedded spheres $S(x), S(y)$. We define the intersection pairing $H_k(M) \times H_k(M) \to \mathbf{Z}$ by the formula

$$x \cdot y = [S(x):S(y)].$$

It follows from VI,10.3 that this is well-defined, bilinear, symmetric for k even, and skew-symmetric for k odd.

Let $H^k(M, \partial M) \times H^k(M, \partial M) \to \mathbf{Z}$ be the cup product pairing defined by $(u, v) = \langle u \smile v, o_M \rangle$, where $o_M \in H_{2k}(M, \partial M)$ is a generator. If u, v are the Poincaré duals of $x, y \in H_k(M)$, then $x \cdot y = (u, v)$ by [D,VIII,13.5]. This simplifies the proof of bilinearity of the intersection product and defines it for all $4n$-dimensional manifolds. (It would be possible to avoid here this reference to cup product and work exclusively with intersection pairing; the proof of 3.3 can be modeled on the proof of 4.2.)

To a symmetric pairing there is associated a quadratic form Q, $Q(x) = x \cdot x$. The pairing is said to be even if Q takes only even values. The signature σ of the pairing is the signature of Q, when diagonalized over the real numbers. If $\dim M = 2k$ with k even, then the intersection pairing is symmetric and both notions are well-defined invariants of M; the signature will be denoted $\sigma(M)$.

(3.1) Proposition *Suppose that M^{2k} is a π-manifold and k is even. Then the intersection pairing is unimodular and even, and $S(x)$ has a trivial normal bundle if and only if $x \cdot x = 0$.*

Proof The pairing is unimodular by VI,12.2. Now, by VI,12.4, with an appropriate identification $\pi_{k-1}(S^{k-1}) \simeq \mathbf{Z}$, we have

$$x \cdot x = \phi_* \alpha(x),$$

where ϕ is the projection of the fibration $SO(k)/SO(k-1)$. We use now the notation and results of A,5. Since M is a π-manifold, the normal bundle of $S(x)$ is stably trivial; thus $\alpha(x) \in \operatorname{Ker} s_k = \operatorname{Im} \partial$. Since for k even ϕ_* is a monomorphism on $\operatorname{Im} \partial$, the last part of 3.1 follows. Since $\operatorname{Im} \phi_*$ consists of even integers, the intersection pairing is even. $\square$

We will now discuss the signature. It follows easily from the definition that

(3.2) $\sigma(-M) = -\sigma(M)$, $\sigma(M_1 \# M_2) = \sigma(M_1) + \sigma(M_2)$,

where the connected sum is taken along the boundary if M_1 and M_2 are bounded.

(3.3) Proposition $\sigma(M)$ *is an invariant of cobordism.*

Proof If M is closed, then this follows from [D,VIII,9.6], since we can identify the cup product pairing and the intersection pairing. If ∂M is a homotopy sphere, then attaching to M a cone on ∂M produces a homology manifold (indeed a topological manifold by VIII,4.6) with the same signature, and the result follows as before. $\square$

We now have all we need to decide when it is possible to perform surgery on $H_k(M)$ if k is even, $k = 2n$.

(3.4) Theorem *Let M be a framed $4n$-dimensional manifold, $n > 1, \partial M$ either empty or a homotopy sphere. There is a framed cobordism between M and a $2n$-connected manifold M' if and only if $\sigma(M) = 0$.*

(Recall the convention of 2.1: *framed manifold* refers to a framing of the stable tangent bundle.)

If M is closed, then by the Signature Theorem of Hirzebruch $\sigma(M) = 0$. Since a $2n$-connected closed $4n$-dimensional manifold is a homotopy sphere, we obtain the following:

(3.5) Corollary *A framed closed $4n$-dimensional manifold, $n > 1$, is framed cobordant to a homotopy sphere.* $\square$

By IX,1.6 this implies that every element of the stable homotopy group $\pi_{4n}(S)$ can be represented (via the Pontriagin construction) by a framed homotopy sphere.

If $\partial M'$ ($= \partial M$) is a homotopy sphere and M' is $2n$-connected, $n > 1$, then M' is diffeomorphic to D^{4n}, *cf.* VIII,4.5. Thus:

(3.6) Corollary *If a homotopy sphere Σ bounds a π-manifold M and $\sigma(M) = 0$, then Σ is diffeomorphic to S^{4n-1}.* $\square$

Proof of 3.4 By 3.3 the vanishing of signature is a necessary condition. It remains to prove the sufficiency and, in view of 2.2, we may assume that M is $(2n - 1)$-connected, $\sigma(M) = 0$.

Consider the intersection pairing. Recall that a symplectic base for it is a base $\mathbf{e}_1, \ldots, \mathbf{e}_t, \mathbf{f}_1, \ldots, \mathbf{f}_t$ on $H_k(M)$ satisfying

$$\mathbf{e}_i \cdot \mathbf{e}_j = \mathbf{f}_i \cdot \mathbf{f}_j = 0, \qquad \mathbf{e}_i \cdot \mathbf{f}_j = \delta_{ij}.$$

A fundamental theorem in the theory of quadratic forms asserts that a symmetric unimodular even pairing admits a symplectic base if and only if it has signature zero (*cf.* [Se,V,Th. 5]). By 3.1 this is our case. Let then $\mathbf{e}_1, \ldots, \mathbf{e}_t, \mathbf{f}_1, \ldots, \mathbf{f}_t$ be a symplectic base for the intersection pairing and let $S = S(\mathbf{e}_1)$ be an imbedded sphere representing $\mathbf{e}_1$. By 3.1 the normal bundle of S is trivial; hence we can perform a surgery on S and, by 2.1, this surgery can be framed. This implies that $\chi(M, S)$ is again a π-manifold. Since $\mathbf{e}_1 \cdot \mathbf{f}_1 = 1$, S is primitive, as this notion was defined in Section 1. By 1.3 $\chi(M, S)$ is $(2n - 1)$-connected, and the rank of $H_{2n}(\chi(M, S))$ is smaller than the rank of $H_{2n}(M)$. Thus a finite sequence of surgeries will reduce it to zero. $\square$

The fact that S is primitive is easily visualized. If $\partial M \neq \emptyset$, then, as noted at the beginning of this section, M is a handlebody and S may be taken as a core of a handle. Then a meridian of S lies in ∂M; hence it bounds there. If M is closed, then M with the interior of a disc removed is a handlebody and the same argument still works.

There is another interesting consequence of the proof of 3.4. We have shown that if M is closed and $(k - 1)$-connected, $k = 2n$, then there is a sequence of surgeries on k-spheres leading to a homotopy sphere Σ. Each surgery amounts to attaching a $(k + 1)$-handle to $M \times I$, and the framed manifold W that realizes the cobordism between M and Σ is of the form

$$W = M \times I \cup ((k + 1)\text{-handles}), \qquad \partial_+ W = \Sigma.$$

The dual presentation is thus $W = \Sigma \times I \cup (k\text{-handles})$ and, according to VI,11.3,

$$W = (\Sigma \times I) \#_b T_1 \#_b T_2 \#_b \cdots \#_b T_s,$$

where $T_1, \ldots, T_s$ are $(k + 1)$-disc bundles over S^k. Since W is a π-manifold, they must be trivial. This means that the two components of ∂W are, respectively, Σ and a connected sum of Σ with a connected sum of a certain number of copies of $S^k \times S^k$. Since this second component is M, we have proved the following:

(3.7) Proposition *A closed $(k - 1)$-connected π-manifold M^{2k}, k even > 2, is diffeomorphic to a connected sum of a homotopy sphere Σ and g copies of $S^k \times S^k$, $g = \frac{1}{2} b_k(M)$.* $\square$

According to [Ko, 3.1], if $M \# \Sigma$ is diffeomorphic to M, M a $(k - 1)$-connected π-manifold, $\Sigma \in \theta^{2k}$, then $\Sigma \simeq S^{2k}$. Thus assigning to a natural number g and a homotopy sphere $\Sigma \in \theta^{2k}$ the connected sum of Σ and g copies of $S^k \times S^k$ establishes a bijective correspondence between the set of all such pairs (g, Σ) and the set of all smooth structures on $(k - 1)$-connected, $2k$-dimensional π-manifolds.

Exercise Two manifolds M, N are said to be almost diffeomorphic if $M - \{p\}$ is diffeomorphic to $N - \{q\}$, $p \in M$, $q \in N$. Show that M is almost diffeomorphic to N if and only if $M \# \Sigma$ is diffeomorphic to N for some homotopy sphere Σ.

Proposition 3.7 asserts that $(k - 1)$-connected, $2k$-dimensional π-manifolds are almost diffeomorphic if and only if they have the same kth Betti number.

All arguments in this section fail completely in the case of 4-dimensional manifolds: It is not, in general, possible to represent a 2-dimensional homology class by an imbedded sphere, cf. [KM1].

4   Surgery on $(4n + 2)$-Dimensional Manifolds

We will now consider a $(k - 1)$-connected π-manifold M of dimension $2k$, where k is odd, $k = 2n + 1$, $n > 0$. The boundary of M is assumed to be either empty or a homotopy sphere. The intersection pairing is now skew-symmetric and unimodular; hence it always admits a symplectic basis, cf. [N,IV.1]. However, 3.1 no longer holds and we need another way of finding spheres with a trivial normal bundle. Our strategy will be based on the following proposition.

(4.1)   Proposition $\alpha(x + y) = \alpha(x) + \alpha(y) + (x \cdot y)\tau_k$.

Note that since M is a π-manifold, $\alpha(x) \in \text{Im } \partial$ and is of order 2 by A,5.

Proof   We will need some results from the theory of immersions, cf. [Wi3] and [M7].

Observe first that if S is an immersed sphere, then it has a well-defined normal bundle (III,2.1) and its characteristic element $\alpha(S)$ is an invariant of regular homotopy of S, i.e., homotopy through immersions (III,2.7). It is easy to see that if S_1, S_2 are two immersed spheres in M, and $S_1 \# S_2$ stands for an immersed sphere obtained by joining S_1 and S_2 by a tube, then

$$(*) \qquad\qquad \alpha(S_1 \# S_2) = \alpha(S_1) + \alpha(S_2).$$

Next, for an immersed sphere there is defined a self-intersection number $\beta(S)$, which is an integer mod 2 and equals 0 if and only if S is regularly homotopic to an imbedding. Moreover

$$(**) \qquad \beta(S_1 \# S_2) = [S_1 : S_2] \quad \mod 2.$$

Finally, we will need the fact that there is an immersion $h_0 : S^k \to \mathbf{R}^{2k}$ with the self-intersection number $\beta(h_0) = 1$.

Now, let $S_1 = S_1(x)$, $S_2 = S_2(y)$ be two imbedded spheres. If $x \cdot y$ is even, then $\beta(S_1 \# S_2) = 0$, by $(**)$, and $S_1 \# S_2$ is regularly homotopic to an imbedding S representing $x + y$. Hence, by $(*)$,

$$\alpha(x + y) = \alpha(S) = \alpha(S_1 \# S_2) = \alpha(S_1) + \alpha(S_2) = \alpha(x) + \alpha(y),$$

which proves 4.1 in this case.

Suppose now that $x \cdot y$ is odd; hence $\beta(S_1 \# S_2) = 1$. Take a chart in M disjoint from $S_1 \# S_2$, and view it as $\mathbf{R}^{2k}$. The immersion h_0 becomes now an immersion of S^k in M representing $0 \in H_k(M)$ and with the self-intersection number 1. Since $\beta(S_1 \# S_2 \# h_0) = 0$, $S_1 \# S_2 \# h_0$ is regularly homotopic to an imbedding representing $x + y$ and, as before,

$$(***) \quad \alpha(x + y) = \alpha(S) = \alpha(S_1 \# S_2 \# h_0)$$
$$= \alpha(S_1) + \alpha(S_2) + \alpha(h_0) = \alpha(x) + \alpha(y) + \alpha(h_0).$$

There remains to be shown that $\alpha(h_0) = \tau_k$. To see this, let $M = S^k \times S^k$, let $S_1 = S^k \times \{p\}$ represent $x \in H_k(S^k \times S^k)$, and let $S_2 = \{p\} \times S^k$ represent y. Then $x + y$ is represented by the diagonal imbedding Δ, and, since $\alpha(x) = \alpha(y) = 0$, we obtain from $(***)$ that $\alpha(\Delta) = \alpha(h_0)$. But it is well-known that $\alpha(\Delta) = \tau_k$. $\quad \square$

(This proof follows essentially Levine's modification of Wall's argument, *cf.* [Le2], [W2].)

It follows from 4.1 that if the rank of $H_k(M)$ is at least four, then there is a symplectic basis for it with at least one element of the basis represented by a sphere with a trivial normal bundle. For suppose that $\mathbf{e}_1, \ldots, \mathbf{e}_t, \mathbf{f}_1, \ldots, \mathbf{f}_t$ is a symplectic basis and that $\mathbf{e}_1, \mathbf{e}_2, \mathbf{f}_1, \mathbf{f}_2$ all have non-trivial normal bundles. Introduce a new basis by the formulae

$$\mathbf{e}_1' = \mathbf{e}_1 + \mathbf{e}_2, \qquad\qquad \mathbf{f}_1' = \mathbf{f}_1 + \mathbf{e}_2,$$
$$\mathbf{e}_2' = \mathbf{e}_1 + \mathbf{e}_2 + \mathbf{f}_1 + \mathbf{f}_2, \qquad \mathbf{f}_2' = \mathbf{f}_1 + \mathbf{f}_2 + \mathbf{e}_1,$$
$$\mathbf{e}_i' = \mathbf{e}_i, \qquad\qquad\qquad \mathbf{f}_i' = \mathbf{f}_i, \qquad i > 2.$$

A calculation shows that $\alpha(\mathbf{e}_1') = \alpha(\mathbf{f}_1') = \alpha(\mathbf{e}_2') = \alpha(\mathbf{f}_2') = 0 \mod 2$.

It is clear now that if the rank of $H_{2n+1}(M)$ is at least four, then we can proceed with surgery in the same way as we did in the proof of 3.4. First, we find a symplectic basis such that $\alpha(e_1) = 0$. The surgery on $S(e_1)$ will reduce the rank by 2 and, by 2.1, it can be framed unless $n = 1$ or $n = 3$. This is no impediment to an inductive procedure, for in these two cases $(2n + 1)$-dimensional spheres in M have trivial normal bundles and we do not need to verify at every stage that we still have a π-manifold. Continuing in this way, we will eventually end up with either a $(2n + 1)$-connected manifold or with a manifold M' with H_{2n+1} of rank 2 and $\alpha(e_1) = \alpha(f_1) = 1$ for a symplectic basis e_1, f_1. (Clearly, the second case cannot happen if $n = 1, 3$.) We will now show that there is an invariant allowing us to decide which of the two cases occurs.

We assume now that $k \neq 3, 7$ and identify $\mathrm{Im}\,\partial$ with Z_2. The bilinear pairing $x \cdot y$ on $H_k(M)$ induces a symmetric bilinear pairing (x, y) to Z_2 by the formula $(x, y) = x \cdot y \bmod 2$. With this notation, the map α satisfies

$$\alpha(x + y) = \alpha(x) + \alpha(y) + (x, y).$$

A function $\alpha: H_k(M) \to Z_2$ such that $\alpha(x + y) - \alpha(x) - \alpha(y)$ is a nonsingular bilinear symmetric pairing is called a quadratic form. Given a symplectic basis $e_1, \ldots, e_t, f_1, \ldots, f_t$ of $H_k(M)$, the Arf invariant of this form is defined to be the number mod 2,

$$\kappa = \alpha(e_1)\alpha(f_1) + \cdots + \alpha(e_t)\alpha(f_t).$$

It is known that κ does not depend on the choice of the symplectic basis, cf. [Br1,III.1]. Therefore it is an invariant of M, the Kervaire invariant $\kappa(M)$.

As an example, note that for the Kervaire manifold $K(4n + 2)$ of VI,12 we have $\kappa(K(4k + 2)) = 1$. Of course, $K(4k + 2)$ is not closed; we will discuss in Section 6 the important problem of the existence of closed manifolds with Kervaire invariant 1. For the closed manifold $S^{2n+1} \times S^{2n+1}$ we have Kervaire invariant equal to zero.

It follows easily from the definition that

(4.2) $\kappa(M_1 \# M_2) = \kappa(M_1) + \kappa(M_2),$

where the connected sum is taken along the boundary if M_1 and M_2 are bounded. (The symplectic bases for M_1 and M_2 yield together a symplectic basis for $M_1 \# M_2$.)

Observe that the Kervaire invariant was defined for all $2n$-connected $(4n + 2)$-dimensional π-manifolds, $n \neq 1, 3$, closed or bounded by a homotopy sphere, and that the framing did not intervene in the definition.

The following proposition will allow us to define it for all framed $(4n + 2)$-dimensional π-manifolds.

(4.3) Proposition *Suppose that* $\{M_0, W, M_1\}$ *is a framed cobordism between two $2n$-connected $(4n + 2)$-dimensional manifolds M_0 and M_1, $n > 1$, $n \neq 3$. Then $\kappa(M_0) = \kappa(M_1)$.*

Proof Since W is a $(4n + 3)$-dimensional π-manifold, we can assume by 2.2 that it is $2n$-connected. This being the case, we conclude from VIII,6.5 that $M_0 \# S_0$ and $M_1 \# S_1$ are diffeomorphic, where S_0 and S_1 are each a connected sum of a number of copies of $S^{2n+1} \times S^{2n+1}$. Therefore

$$\kappa(M_0) = \kappa(M_0 \# S_0) = \kappa(M_1 \# S_1) = \kappa(M_1). \qquad \square$$

We assumed in 4.3 that M_0 and M_1 are closed manifolds. If they are bounded by homotopy spheres and cobordant in the sense of Section 1, then there exists a framed manifold W such that $\partial W = M_0 \cup (\partial M_0 \times I) \cup M_1$. By 4.3 we have $\kappa(\partial W) = 0$. Since we can find a symplectic basis for $H_{2n+1}(\partial W)$ that splits into symplectic bases for $H_{2n+1}(M_0)$ and $H_{2n+1}(M_1)$, we obtain again $\kappa(M_0) = \kappa(M_1)$.

It follows from these arguments that if $(M, \mathbf{F})$ is a $(4n + 2)$-dimensional framed π-manifold, closed or bounded by a homotopy sphere, then we can define its Kervaire invariant $\kappa(M, \mathbf{F})$ as $\kappa(M')$, where M' is $2n$-connected and framed cobordant to $(M, \mathbf{F})$. The existence of such M' is guaranteed by 2.2. The relation 4.2 becomes

(4.4) $$\kappa(M_1 \# M_2, \mathbf{G}) = \kappa(M_1, \mathbf{F}_1) + \kappa(M_2, \mathbf{F}_2)$$

for an appropriately chosen framing $\mathbf{G}$ (*cf.* IX,1.6 and 2.4). We now collect the results of this section.

(4.5) Theorem *Let $(M, \mathbf{F})$ be a $(4n + 2)$-dimensional framed π-manifold, closed or bounded by a homotopy sphere.*

If $n \neq 1, 3$, then there is a framed cobordism between M and a $(2n + 1)$-connected manifold M' if and only if $\kappa(M, \mathbf{F}) = 0$.

If $n = 1, 3$, then such a cobordism, not necessarily framed, always exists. $\square$

(4.6) Corollary *If a $(4n + 1)$-dimensional homotopy sphere Σ bounds a π-manifold M with $\kappa(M) = 0$, then Σ is diffeomorphic to S^{4n+1}.*

Proof By 4.5 Σ bounds a contractible manifold; thus this follows from VIII,4.5. (If $n = 1, 3$, then κ is not defined and the assumption $\kappa(M) = 0$ is superfluous.) $\square$

(4.7) Corollary *There exists a closed $(4n + 2)$-dimensional π-manifold M with $\kappa(M) = 1$ if and only if the boundary of the Kervaire manifold $K(4n + 2)$ is diffeomorphic to S^{4n+1}.*

Proof Let M be such a manifold, let N be M with a disc removed, and let $W = N \#_b K(4n + 2)$. Then $\partial W = S^{4n+1}$ by 4.2 and 4.6. But $\partial W = \partial N \# \partial K(4n + 2) = \partial K(4n + 2)$, for $\partial N = S^{4n+1}$. The converse is obvious. $\square$

The following proposition is an analogue of 3.7:

(4.8) Proposition *A closed $(k - 1)$-connected π-manifold M^{2k}, k odd, with $\kappa(M^{2k}) = 0$ is diffeomorphic to a connected sum of a homotopy sphere Σ with g copies of $S^k \times S^k$, $g = \frac{1}{2}b_k(M^{2k})$.*

Proof If $k \neq 3, 7$, then the proof of 3.7 applies without change. If $k = 3$ or 7, then it yields an analogous result but with $S^k \times S^k$ replaced by k-sphere bundles over S^k. Such bundles are classified by $\pi_{k-1}(\mathbf{SO}(k + 1))$, which is trivial for $k = 3$ or 7, *cf.* A,5.1. Thus the bundles in question are actually product bundles. $\square$

Note that, by 4.2, the conclusion of 4.8 always holds for $M^{2k} \# M^{2k}$.

5   Surgery on Odd-Dimensional Manifolds

In contradistinction to the case of even-dimensional manifolds there is no obstruction to the surgery in the middle dimension on odd-dimensional manifolds.

(5.1) Theorem *Let $\{\Sigma_1, W, \Sigma_2\}$ be a framed cobordism between two homotopy spheres Σ_1, Σ_2. If $\dim W = 2k + 1 > 3$, then W is framed cobordant to an h-cobordism.*

In the following two corollaries M stands for a framed $(2k + 1)$-dimensional manifold, $k > 1$.

(5.2)   Corollary   *If M is closed, then it is framed cobordant to a homotopy sphere.*   □

(5.3)   Corollary   *If $\partial M = \Sigma$ is a homotopy sphere, then Σ bounds a contractible manifold.*   □

It follows that Σ is diffeomorphic to S^{2k} if $k > 2$.

Proof of 5.1   By 2.2 we can assume that W is $(k-1)$-connected and we have to prove that $H_k(W)$ can be killed by a sequence of framed surgeries. The proof will be based on the description of W as $W_1 \cup_h W_2$, where

$$W_1 = (\Sigma_1 \times I) \#_b T_1 \#_b \cdots \#_b T_s, \qquad W_2 = (\Sigma_2 \times I) \#_b T_1' \#_b \cdots \#_b T_s',$$

the T_i and T_j' are copies of $S^k \times D^{k+1}$, and h is a diffeomorphism of boundaries. We obtained this representation in IX,7.3 starting with the handle presentation

$$W = \Sigma_1 \times I \cup H_1^k \cup \cdots \cup H_s^k \cup H_1^{k+1} \cup \cdots \cup H_t^{k+1},$$

and applying VIII,6.4 and IX,7.2. In particular, the tori T_i correspond to the k-handles, and the tori T_i' correspond to the $(k+1)$-handles.

To each presentation we will associate two square $s \times s$ matrices $\mathfrak{M}_1$ and $\mathfrak{M}_2$ made of intersection numbers of meridians and equators of the T_j and the T_i'. More precisely, let m_i, m_i' stand for the oriented meridian of T_i and T_i' respectively and let e_j, e_j' be their equators. Then

$$\mathfrak{M}_1 = (b_{ij}), \quad b_{ij} = [m_i : m_j'], \qquad \mathfrak{M}_2 = (c_{ij}), \quad c_{ij} = [e_i : m_j'],$$

$$i, j = 1, \ldots, s.$$

Thus in both matrices rows are indexed by k-handles, i.e., the T_i, and the columns by $(k+1)$-handles, i.e., the T_i'.

Note that m_i can also be viewed as the belt sphere of the ith handle H_i^k, and m_j' as the attaching sphere of H_j^{k+1}; thus $\mathfrak{M}_1$ is the intersection matrix from VII,3, denoted there by $\mathfrak{M}_{k+1}$. It determines completely the homology of W. In particular, if $\mathfrak{M}_1$ is diagonal with ± 1's on the diagonal, then $H_k(W) = 0$.

(5.4)   Lemma   g.c.d. $(b_{1i}, \ldots, b_{si}, c_{1i}, \ldots, c_{si}) = 1$, $\qquad i = 1, \ldots, s$.

Proof   We view m_i, m_i' as spheres in ∂W_1. Since $[m_i' : e_i'] = \pm 1$, m_i' represents a primitive homology class in $H_k(\partial W_1)$. Since $H_k(\partial W_1)$ is generated by the

e_j and the m_j, we have, for some integers x_j, y_j,

$$m'_i = \sum_j (b_{ji}e_j + c_{ji}m_j), \qquad e'_i = \sum_j (x_je_j + y_jm_j).$$

(We are using the same letter to stand for spheres and homology classes they represent.) Thus, with the dot for the intersection product,

$$m'_i \cdot e'_i = \sum_j b_{ji}y_j + \sum_j c_{ji}x_j = \pm 1,$$

for $e_i \cdot e_j = 0 = m_i \cdot m_j$, $e_i \cdot m_j = \pm \delta_{ij}$. $\qquad \square$

The notion of the equator e_i depends on the representation of T_i as the product $S^k \times D^{k+1}$. Let $m = S^k \times \{p\}$, $e = \{q\} \times \partial D^{k+1}$ be a meridian and an equator of $S^k \times D^{k+1}$, and let $\gamma: S^k \to \mathbf{SO}(k+1)$. The diffeomorphism h_γ, $h_\gamma(x, y) = (x, \gamma(x) \cdot y)$ yields a new parametrization of $S^k \times D^{k+1}$ with the new meridian $e_\gamma = h_\gamma(e)$. The homology classes of e and e_γ in $\partial(S^k \times D^{k+1})$ are related by

$$[e_\gamma] = [e] + \phi_*[\gamma][m],$$

where $\phi_*: \pi_k(\mathbf{SO}(k+1)) \to \pi_k(S^k)$ is from the homotopy exact sequence of the fibration $\mathbf{SO}(k+1)/\mathbf{SO}(k) = S^k$ and we have identified $\pi_k(S^k)$ with $\mathbf{Z}$, cf. A,5.

This yields

$$[e_\gamma : S] = [e : S] + \phi_*[\gamma][m : S]$$

for any closed and oriented k-dimensional submanifold S of $\partial(S^k \times D^{k+1})$.

It follows that the effect of a reparametrization of T_i by $[\gamma] \in \pi_k(\mathbf{SO}(k+1))$ is to replace the ith row of $\mathfrak{M}_2$ by its sum with the ith row of $\mathfrak{M}_1$ multiplied by $\phi_*[\gamma]$; $\mathfrak{M}_1$ remains unchanged.

Suppose now that we perform surgery on the core of the ith torus T_i; this is possible since its core is a k-sphere with a trivial normal bundle. We will call a surgery on it a surgery on the ith handle. Viewing a surgery as a replacement of $S^k \times D^{k+1}$ by $D^{k+1} \times S^k$, we see that the effect of a surgery on the ith handle is the interchange of the ith rows of $\mathfrak{M}_1$ and $\mathfrak{M}_2$. However, if we want surgery to be framed, then before pasting $D^{k+1} \times S^k$ in, we may have to twist it by the map $(x, y) \mapsto (\gamma(y) \cdot x, y)$, where the homotopy class of $s_*[\gamma]$ in $\pi_k(\mathbf{SO}(k+2))$ is determined by the obstruction to framing, cf. 2.1. This is equivalent to a reparametrization of T_i by h_γ before performing the replacement of $S^k \times D^{k+1}$ by $D^{k+1} \times S^k$. Combining this with what we know about the effects of reparametrization, we obtain the following lemma.

(5.5) Lemma *The effect of a framed surgery on T_i on the first columns of* $\mathfrak{M}_1$ *and* $\mathfrak{M}_2$ *is to substitute* $c_{i1} + \phi_*(\gamma)b_{i1}$ *for* b_{i1}, *and* b_{i1} *for* c_{i1}.   □

We proceed now with the proof of 5.1. Assume that W has a presentation with s k-handles. We will show that if $s > 0$, then there is a sequence of surgeries on W leading to a manifold W' such that the first column of $\mathfrak{M}_1$ has relatively prime entries; such a sequence of surgeries will be called handle reducing. This will prove 5.1, for in this case there is a sequence of elementary operations on rows and columns of $\mathfrak{M}_1$ resulting in a matrix in which first row and first column intersect in ± 1 and have zeros elsewhere. An inductive argument shows then that we can eventually obtain $\mathfrak{M}_1$ with diagonal elements equal to ± 1 and zeros elsewhere. But then $H_k(W') = 0$ and W' is an h-cobordism. (Alternately, if $k > 2$, we can apply VIII,2.3 and conclude that there is a presentation of W' with $s - 1$ k-handles.)

Suppose now that the presentation of W satisfies the following condition:

(∗) $b_{21} = \cdots = b_{s1} = 0$ and either $b_{11}c_{11} = 0$ or $|b_{11}| = |c_{11}|$.

We claim that in this case W admits a handle-reducing surgery. For in either case a surgery on all T_i such that $b_{i1} = 0$ will result, by 5.5, in the matrix $\mathfrak{M}_1$ with relatively prime entries in the first column.

The proof of 5.1 will now be concluded by showing that if $s > 0$, then W admits a presentation satisfying (∗). We will consider first the case of k even, > 2.

Certainly, W admits a presentation with $\mathfrak{M}_1$ diagonal (*cf.* VIII,1.5). By VIII,2.1 we can assume that the meridian m_1' does not intersect meridians of $T_2, \ldots, T_s$; thus it can be viewed as lying in the boundary of $(\Sigma_1 \times I) \#_b T_1$, i.e., in $\Sigma_1 \# (S^k \times S^k)$; we can assume it misses Σ_1. It has a trivial normal bundle—it is the attaching sphere of the first $(k + 1)$-handle—and therefore it can be framed and the Pontriagin construction will then yield a map $S^k \times S^k \to S^k$ of bidegree (b_{11}, c_{11}). It is well-known (*cf.* [SE,I, § 5]) that this implies the existence of a map $S^{2k+1} \to S^{k+1}$ with Hopf invariant $b_{11}c_{11}$. But if $k + 1$ is odd, then the Hopf invariant of such a map equals zero. This shows that a presentation of W with $\mathfrak{M}_1$ diagonal must satisfy (∗) and concludes the proof of 5.1 for k even, > 2.

If $k = 2$, i.e., dim $W = 5$, we cannot use VIII,2.1. To get the same conclusion we apply instead the following lemma, in which T is a boundary connected sum of s copies of $S^k \times D^{k+1}$ and $S \subset \partial T$ is a k-sphere representing the class $\sum_i (b_i m_i + c_i e_i) \in H_k(\partial T)$; $m_i, e_i, i = 1, \ldots, s$, are homology classes of meridians and equators of ∂T.

Lemma *There is an imbedding of T in S^{2k+1} and a normal framing $\mathbf{F}$ of S such that the map $p(S, \mathbf{F})$: $S^{2k+1} \to S^{k+1}$ has Hopf invariant $\sum_i b_i c_i$.*

This lemma is easily proved using the definition of the Hopf invariant as the linking number of pre-images. It is left as an exercise for readers familiar with it. (Choose the imbedding so that the e_i bound in the complement of T.)

We will now prove 5.1 for the case when k is an odd integer.

We begin with a presentation of W satisfying two conditions:

(a) $\mathfrak{M}_1$ is diagonal;
(b) A framed surgery on T_1 is possible without further reparametrization. Observe that:
(c) If (b) holds and we perform a framed surgery on T_1 replacing it by $\bar{T}_1$ then (b) holds for $\bar{T}_1$.

(This is essentially a tautology.)

We will prove by induction on $|b_{11}|$ that W admits a handle-reducing surgery. By ($*$) this is true if either $b_{11}c_{11} = 0$ or $|b_{11}| = |c_{11}|$. We assume then that neither holds and consider two cases.

Case 1: $0 < |c_{11}| < |b_{11}|$. In this case there is an integer m such that either $|b_{11} - 2mc_{11}|$ or $|b_{11} - (2m + 2)c_{11}|$ is $\leq |c_{11}|$; say $|b_{11} - 2mc_{11}| \leq |c_{11}|$. Perform a framed surgery on T_1; this will interchange c_{11} and b_{11} and preserve (b), *cf.* 5.5 and (c) preceding. Reparametrize the new first handle using $\gamma \in \operatorname{Im} \partial$ chosen so that $\phi_*(\gamma) = -2m$; this is possible by A,5.2(b). (We now use the notation of A,5 with the index shift by 1.) This will not affect (b), for $s_{k+1}(\gamma) = 0$. Therefore we can again perform a surgery which will result in the element b'_{11} on top of the first column of $\mathfrak{M}_1$ equal to $|b_{11} - 2mc_{11}|$. Since

$$|b'_{11}| = |b_{11} - 2mc_{11}| \leq |c_{11}| < |b_{11}|,$$

we have succeeded in decreasing $|b_{11}|$.

Case 2: $0 < |b_{11}| < |c_{11}|$. In this case there is an integer m such that either $|c_{11} - 2mb_{11}|$ or $|c_{11} - (2m + 2)b_{11}|$ is $\leq |b_{11}|$; say $|c_{11} - 2mb_{11}| \leq |b_{11}|$. Reparametrize T_1 using $\gamma \in \operatorname{Im} \partial$ chosen so that $\phi_*(\gamma) = -2m$; this is possible by A,5.2(b). Since $s_{k+1}(\gamma) = 0$, (b) is preserved and the new element c'_{11} at the top of the first column of $\mathfrak{M}_2$ satisfies

$$0 \leq |c'_{11}| = |c_{11} - 2mb_{11}| \leq |b_{11}|.$$

If either $c'_{11} = 0$ or $|c'_{11}| = |b_{11}|$ then, by $(*)$, we are done. If neither holds, then we have reduced case 2 to case 1.

The proof of 5.1 is now complete. $\square$

It may be worth noting that if $k \neq 3, 7$, then the argument used to prove 5.1 for k odd can be greatly simplified using A,5.2(c).

(5.6) Proposition *If $M^{2k+1}, k > 1$, is a closed $(k-1)$-connected π-manifold, then, for some homotopy sphere Σ, $M^{2k+1} \# \Sigma$ is the boundary of a parallelizable $(2k+2, k+1)$-handlebody.*

Proof There is a framed cobordism between M and Σ involving only $(k+1)$-handles, hence the same is true for $M \# \Sigma$ and S^{2k+1}. Now, the argument used to prove 3.7 easily adapts to yield 5.6. $\square$

A classification of such handlebodies was undertaken by Wall in [W2].

6 Computation of θ^n

We now apply the results of the last three sections to the computation of the group θ^m, $m \geq 4$. For this purpose we introduce two groups: the group P^m of oriented framed manifolds bounded by a homotopy sphere, and its subgroup P_0^m of manifolds with the boundary diffeomorphic to S^{m-1}. The equivalence relation is given by framed cobordism (as defined in Section 1) and the group operation by the connected sum along the boundary; the unit element is the framed disc D^m. If $m > 5$ then P^m and P_0^m actually are groups, not just commutative monoids: For m even $(-M, \mathbf{F})$ is the inverse of $(M, \mathbf{F})$ by 3.2, 3.4, 4.4 and 4.5; for m odd $P^m = 1$ by 5.1.

Consider the sequence

$$(6.1) \qquad\qquad 0 \to P_0^{m+1} \to P^{m+1} \xrightarrow{b} \theta^m,$$

where the boundary homomorphism b is given by taking the boundary and forgetting the framing. Its image is traditionally denoted bP^{m+1}. This sequence is exact for $m \geq 5$. For $W^{m+1} \in \operatorname{Ker} b$ means that ∂W bounds a contractible manifold and is thus diffeomorphic to S^m by VIII,4.5.

(6.2)   Proposition bP^{m+1} *is a finite cyclic group of order*:

(a) $t_n/8$, t_n *as in* IX,8.7, *if* $m = 4n - 1$ *and* $n > 1$;

(b) 1 *or* 2 *if* $m = 4n + 1$ *and* $n \geq 1$;

(c) 1 *if* $m = 5, 13$;

(d) 1 *if* $m = 2n \geq 4$.

Proof   Let $m = 4n - 1$, $n > 1$. By 3.2 and 3.3, assigning to an element of P^{4n} its signature defines a homomorphism $\sigma: P^{4n} \to \mathbf{Z}$. By 3.4 it is a monomorphism; we claim that its image is the subgroup $8\mathbf{Z}$ of multiples of 8.

To see this, we observe first that the signature of an element of P^{4n} is divisible by 8. This follows from [Se,V, § 2] since the matrix of the intersection pairing is unimodular and even by 3.1. It remains to notice that there is a parallelizable manifold of dimension $4n$ bounded by a homotopy sphere and with signature equal to 8: by IX,7.5 this is the manifold $M(4n)$ constructed in VI,12. It follows that σ is an isomorphism $P^{4n} \simeq 8\mathbf{Z}$.

Now, the image of σ restricted to P_0^{4n}, as already computed in IX,8.5, is precisely $t_n\mathbf{Z}$. Therefore (a) follows from the exact sequence 6.1. Observe that $\partial M(4n)$ is the generator of bP^{4n}.

To calculate bP^{4n+2} we employ the homomorphism $\kappa: P^{4n+2} \to \mathbf{Z}_2$ defined by assigning to $(M, \mathbf{F})$ its Kervaire invariant $\kappa(M, \mathbf{F})$. By 4.3 and 4.4 κ is a well-defined homomorphism if $n \neq 1, 3$; by 4.5 it is a monomorphism. It is an isomorphism, for we have constructed in VI,12 a parallelizable manifold $K(4n + 2)$ bounded by a homotopy sphere and observed in Section 4 that $\kappa(K(4n + 2)) = 1$. Together with 6.1 this proves (b) for $n \neq 1, 3$.

Now, (c) follows from 4.6 and (d) follows from 5.3. $\square$

Since $t_2/8 = 28$, it follows that $bP^8 = \mathbf{Z}_{28}$, generated by $\partial M(8)$. In particular, the 7-dimensional homotopy sphere $\partial M(8)$ is homeomorphic but not diffeomorphic to S^7. Thus we have obtained the first example here of a nonstandard smooth structure on a sphere. Historically, the first example was due to J. Milnor who showed in 1956 [M3] that S^7 admits at least 7 distinct differentiable structures. This unexpected result attracted great attention and gave a powerful stimulus to the development of differential topology.

Let Σ be a 4- or 5-dimensional homotopy sphere. It is a π-manifold; hence, when imbedded in a high-dimensional sphere, its normal bundle will admit a framing $\mathbf{F}$ and $(\Sigma, \mathbf{F})$ will represent an element of the stable homotopy group $\pi_m(S)$. Since $\pi_m(S) = 0$ if $m = 4$ or 5 [To,XIV], $(\Sigma, \mathbf{F})$ bounds a parallelizable manifold. This shows that $\theta^m = bP^{m+1}$ for $m = 4, 5$; hence by 6.2(c) and (d),

$$(6.3) \qquad\qquad \theta^4 = \theta^5 = 0.$$

The argument used in this proof can be refined to extend the sequence 6.1.

Recall that the group Ω_f^m of framed cobordism classes of framed π-manifolds can be identified with the stable homotopy group $\pi_m(S)$ and its subgroup S_f^m of framed spheres with the image of the Hopf–Whitehead homomorphism J_m, cf. IX,5.5 and 6.3.1. Let $\Sigma_f^m \subset \Omega_f^m$ be the group of framed cobordism classes of framed homotopy spheres. We will construct a homomorphism $p\colon \theta^m \to \Sigma_f^m / S_f^m$ such that the sequence

$$(6.4) \qquad\qquad 0 \to bP^{m+1} \to \theta^m \overset{p}{\to} \Sigma_f^m / S_f^m \to 0$$

is exact for $m \geq 5$.

Given $\Sigma \in \theta^m$, we consider the set $p(\Sigma) \subset \Sigma_f^m$ consisting of classes represented by all possible framings of Σ. We claim that $p(\Sigma)$ is a coset of S_f^m. To see this, recall first that, by IX,5.7,

$$(*) \qquad\qquad p(\Sigma, \mathbf{F}) + p(\Sigma', \mathbf{F}') = p(\Sigma \# \Sigma', \mathbf{G})$$

for an appropriately chosen framing $\mathbf{G}$. Letting $\Sigma' = -\Sigma$, we get

$$p(\Sigma, \mathbf{F}) - p(\Sigma, \mathbf{F}') = p(S^m, \mathbf{G})$$

since $\Sigma \#(-\Sigma)$ bounds a contractible manifold and is thus diffeomorphic to S^m by VIII,4.5. This shows that $p(\Sigma)$ is contained in a coset $p(\Sigma, \mathbf{F}) + S_f^m$ of S_f^m.

Every element of this coset is represented by a map $p(\Sigma, \mathbf{F}) + p(S^m, \mathbf{F}')$ for a suitably chosen framing $\mathbf{F}'$. But letting $\Sigma' = S^m$ we get, from $(*)$,

$$p(\Sigma, \mathbf{F}) + p(S^m, \mathbf{F}') = p(\Sigma, \mathbf{G}) \in p(\Sigma);$$

thus $p(\Sigma, \mathbf{F}) + S_f^m = p(\Sigma)$. Consequently, p is a map $\theta^m \to \Sigma_f^m / S_f^m$, and since homotopy spheres are π-manifolds, it is surjective. That it is a homomorphism follows from $(*)$.

Now, suppose that $p(\Sigma) = S_f^m$. Since S_f^m contains a null-homotopic map, there is a framing $\mathbf{F}$ such that $p(\Sigma, \mathbf{F})$ is null-homotopic. This means that $(\Sigma, \mathbf{F})$ bounds a framed manifold, i.e., $\operatorname{Ker} p \subset bP^{m+1}$. Since the inverse inclusion is obvious, the exactness of 6.4 is proved.

We have seen that bP^{m+1} is finite for $m \geq 4$ and, by a well-known theorem of J.-P. Serre, $\Omega_f^m = \pi_m(S)$ is finite (*cf.* [Sp,9.7]). Thus it follows from 6.3 and 6.4 that:

(6.5) Theorem θ^m *is finite for* $m \geq 4$.   □

To obtain more precise results about θ^m, we patch 6.4 and the exact sequence

$$0 \to \Sigma_f^m / S_f^m \to \Omega_f^m / S_f^m = \text{Coker } J_m \to \Omega_f^m / \Sigma_f^m \to 0,$$

and obtain for $m \geq 5$ the exact sequence

(6.6) $0 \to bP^{m+1} \to \theta^m \to \text{Coker } J_m \to \Omega_f^m / \Sigma_f^m \to 0.$

(6.7) Theorem *Let* $m > 4$, $m \neq 6, 14$. *Then,* Ω_f^m / Σ_f^m *is trivial if* $m \neq 2 \bmod 4$ *and of order* ≤ 2 *if* $m = 2 \bmod 4$.

Proof According to IX,1.6 all results of Sections 3–5 are valid with the framings interpreted as framings of the stable normal bundle. It follows that for m odd the theorem is a consequence of 5.2, and for $m \equiv 0 \bmod 4$ of 3.5. If $m = 4n + 2$, $n \neq 1, 3$, then assigning to an element $(M, \mathbf{F}) \in \Omega_f^m$ its Kervaire invariant defines a homomorphism $\kappa \colon \Omega_f^m \to \mathbf{Z}_2$. By 4.5 the kernel of this homomorphism is precisely Σ_f^m. Thus Ω_f^m / Σ_f^m is of order 2 if and only if κ is surjective, i.e., if there exists a closed π-manifold with Kervaire invariant 1.   □

Further calculations of θ^m depend on precise knowledge of Coker J_m and on the resolution of the Kervaire invariant ambiguity if $m \equiv 1, 2 \bmod 4$. Using the tables of $\pi_n(S)$ in [To] we obtain the following results in low dimensions. The order of a group G is denoted $|G|$.

If $m = 4n$, we have from 6.2, 6.6, and 6.7 that $\theta^{4n} \simeq \text{Coker } J_{4n}$. Since Im $J_{4n} = \mathbf{Z}_2$ for n even and is trivial for n odd, a comparison with Toda's tables yields:

$$\theta^8 = \mathbf{Z}_2, \qquad \theta^{12} = 0, \qquad \theta^{16} = \mathbf{Z}_2.$$

If $m = 4n - 1$, sequence 6.6 becomes

$$0 \to bP^{4n} \to \theta^{4n-1} \to \text{Coker } J_{4n-1} \to 0,$$

where bP^{4n} is cyclic of order $t_n/8$. Since Coker $J_7 = $ Coker $J_{11} = 0$, we see that

$$\theta^7 = \mathbf{Z}_{28}, \qquad \theta^{11} = \mathbf{Z}_{992}.$$

To calculate θ^m for $m \equiv 1, 2 \mod 4$ we need to know whether there exists a closed $(4n + 2)$-dimensional manifold with Kervaire invariant 1. As we have seen in 4.7, an equivalent question is whether $K(4n + 2)$ is diffeomorphic to the standard sphere. Leaving aside for a moment the dimensions 2, 6, 14, where we did not define the invariant, the first dimension that occurs is 10. In this case M. Kervaire in the paper [K2], in which he introduced the invariant, showed that $\partial K(10)$ is not diffeomorphic to S^9. Thus $bP^{10} = \mathbf{Z}_2$, $\Omega_f^{10}/\Sigma_f^{10} = 0$, and we obtain, from 6.6,

$$|\theta^9| = 8, \qquad |\theta^{10}| = 6.$$

A considerable amount of work has been expended on the Kervaire invariant problem in higher dimensions. The farthest reaching result is due to W. Browder who proved in [Br2] that $\partial K(4n + 2)$ is not diffeomorphic to the standard sphere unless $n + 1$ is a power of 2 and, in fact, is diffeomorphic to it if $n = 7$.

The cases $n = 0, 1, 3$ need separate discussion. Our definition of the Kervaire invariant excluded corresponding dimensions 2, 6, and 14. However, already Pontriagin [Po2] defined in these dimensions an invariant of framed manifolds, which is an obstruction to framed surgery, and showed that it equals 1 for certain framings of $S^1 \times S^1$, $S^3 \times S^3$, and $S^7 \times S^7$. The Kervaire invariant can be redefined so that the definition will encompass these dimensions and coincide there with Pontriagin's. It follows that $\Omega_f^m/\Sigma_f^m = \mathbf{Z}_2$ for $m = 2, 6, 14$, which implies $\theta^6 = 0$, $\theta^{13} = \mathbf{Z}_3$, $\theta^{14} = \mathbf{Z}_2$.

7   Historical Note

The theory presented in this chapter was developed in the six year period following Milnor's discovery of nonstandard smooth structures on the 7-sphere. This period was characterized by an extraordinary meshing of the results of mathematicians working in diverse parts of topology.

(7.1) The natural problem of classification of smooth structures on S^n faced two difficulties at the outset. First, in order to define rigorously the operation of connected sum one needs the Disc Theorem, III,3.6. This was proved independently by J. Cerf and R. Palais and published in 1960. Second, even with this operation defined, the problem of the existence of the inverse in the monoid of differentiable structures was very hard. Here the solution came in substituting h-cobordism, defined by R. Thom in [T5], for diffeomorphism as the equivalence relation between smooth structures. This shifted the arguments from constructing diffeomorphisms, an impossible task before Smale, to homotopy theory. The stage was set for the construction of the groups θ^n and their computation.

The groups θ^n appeared for the first time in Milnor's notes "Differentiable manifolds which are homotopy spheres," dated January 23, 1959. This paper was widely distributed but never published; we will refer to it here as DM. Its main results were published in [M5] and [M7]. They were:

—the construction of θ^n;
—the proof that homotopy spheres are π-manifolds, IX,8.6; and
—introduction of the method of surgery and its application to the calculation of bP^{4k}.

The proof that n-dimensional homotopy spheres are π-manifolds depended for $n = 4k$ on certain divisibility properties of Pontriagin numbers due to Kervaire (1957), and on the Hirzebruch Signature Theorem of 1956. For $n = 1, 2 \bmod 8$ the argument led to the question whether the J-homomorphism in the corresponding dimensions is injective, and became conclusive with Adam's positive answer two years later.

The technique of surgery below the middle dimension was fully developed in DM; its use was credited to Thom. Surgery in the middle dimension was treated only for $4k$-dimensional manifolds concluding with the determination of the order of bP^{4k} as $t_k/8$ with t_k as here in 6.2(a), and the formula IX,8.7 for t_k already established in [MK]. The final determination needed, again, the results of Adams.

It was announced in DM that by "making use of the Arf invariant of a certain quadratic form" one can show that the order of bP^{4k+2} is at most 2. No further details were given. The case of even dimensional spheres, i.e., of bP^{2k+1}, was posed as a problem.

By the end of 1959 the results of Smale elucidated the relationship between h-cobordism and diffeomorphism. The group θ^n, $n > 4$, turned out to be, after all, the group of smooth structures on S^n, cf. VIII,5.6. The proof that $bP^{2k+1} = 0$ was provided by Kervaire and Milnor in [KM2] and, independently, by Wall in [W1]. (Wall's paper was submitted earlier, July 21, 1961, but it acknowledges that Kervaire and Milnor have obtained the same result.)

The publication of [KM2] in 1962 filled the remaining gap by providing a construction of the Kervaire invariant. Since the results of Adams became available, the calculation of θ^n was complete—except for the troublesome question of the Kervaire invariant. "Complete," of course, in the mathematical sense; that is, reduced to another unsolved problem, that of the determination of homotopy groups of spheres.

Actually, [KM2] omitted some calculations and constructions. They were relegated to Part II, which was never published. A number of people attempted to reconstitute it in courses and seminars; notes from J. Levine's course were published in [Le2] (beware of misprints!).

(7.2) Even before [KM2] was published A. Haefliger [H2] applied the technique of surgery to obtain an example of differentiably knotted spheres in Euclidean space with codimension greater than 2. This was unexpected, for shortly before Zeeman had shown that in the combinatorial case knotting can take place only in codimension 2. Haefliger extended the notion of h-cobordism and of the connected sum to obtain the group $\Sigma^{m,n}$ of h-cobordism classes of imbeddings $S^n \to S^m$, and calculated $\Sigma^{6k,4k-1}$. In turn, J. Levine constructed a larger group $\theta^{m,n}$ of homotopy n-spheres in S^m, which provided the additional information on which homotopy n-spheres imbed in S^m, [Le1]. This group was studied using a sophisticated version of surgery techniques applied to submanifolds of S^m. The final results appeared as an interrelated family of exact sequences. Since for m large $\theta^{m,n}$ becomes isomorphic to θ^n, Levine's sequences contained the results of [KM2] and could be viewed as their unstable generalization.

This direction of research was continued by Haefliger, Kervaire, and Levine.

Another direction was initiated in 1962 by W. Browder and S. Novikov. Their point of view can be, with some simplifications, stated as follows. We are given manifolds M and X, a k-vector bundle ξ over X, a map $f: M \to X$ of degree 1 and a bundle map $b: \nu \to \xi$ covering f; ν is a normal bundle of some imbedding of M in the Euclidean space of high dimension. The problem is to decide whether there is a cobordism between this configuration and one in which f is a homotopy equivalence. (This is Browder's version of the problem, *cf.* [Br1].)

If X is a single point, then we have, of course, the case of surgery on π-manifolds and the work of Kervaire and Milnor can be viewed as the determination of the obstruction to obtain the desired cobordism through a sequence of surgeries. It turns out that in the general case there is also a well defined obstruction that can be calculated if X is simply connected.

The principal application, indeed, the original motivation, is to classify (up to a diffeomorphism or almost diffeomorphism) the set of manifolds of the same homotopy type. A comprehensive treatment of this subject was given by Browder in [Br1].

Without the assumption of simple connectivity of X, the theory becomes considerably more difficult. This research was initiated by Wall and early results can be found in [W3].

The method of surgery was applied successfully in the theory of imbeddings and in the investigation of group actions on manifolds. (An introduction to the use of surgery in the latter subject can be found in [PR].) At present there are no comprehensive surveys of these and other applications.

(7.3) The homomorphism $\theta^m \to \text{Coker } J_m$ of 6.6 does not provide a method for an explicit construction of exotic smooth structures on spheres. The first such examples were provided by 3-sphere bundles over S^4. In [M4] Milnor introduced the operation of plumbing disc bundles over spheres (cf. VI,12), and obtained a new large class of explicit examples of exotic smooth structures on $(4k - 1)$-spheres. As we have seen in 6.2 one can obtain in this way generators of bP^{2k}. A special case of plumbing of two disc bundles yields a bilinear pairing of the subgroup $s_* \pi_{k-1}(\text{SO}(k - 1))$ of $\pi_{k-1}(\text{SO}(k))$ with itself to θ^{2k-1}, with a large image if $k = 0 \bmod 8$, cf. [Ko].

In 1966, E. Brieskorn discovered a new class of explicit examples of exotic smooth structures. The remarkable feature of his examples is that they occur in a rather classical context: as boundaries of small neighborhoods of an isolated singularity of an affine variety. More precisely, let V_a be the complex hypersurface

$$(z_1)^{a_1} + (z_2)^{a_2} + \cdots + (z_{k+1})^{a_{k+1}} = 0$$

in $\mathbb{C}^{k+1}$, $a = (a_1, a_2, \ldots, a_{k+1})$. Then, with an appropriate choice of a, the intersection of V_a with a small sphere centered at 0 is a smooth $(2k - 1)$-dimensional homotopy sphere and all elements of bP^{2k} can be obtained in this way, cf. [Bk] and [M9]. In particular, all these spheres can be imbedded in the Euclidean space with codimension 2.

Plumbing and the Brieskorn construction seem still to be the only known methods for an explicit construction of homotopy spheres.

Appendix I

In Sections 1 and 2, we present some consequences of the Implicit Function Theorem important in the study of smooth manifolds but difficult to find in textbooks. Section 3 contains the Sard-Brown Theorem in a form adapted to our purposes. In Section 4 we discuss orthonormalization procedures, with an emphasis on the uniqueness and smoothness of the resulting decomposition of matrices. In Section 5 we collect various calculations of the homotopy groups of the orthogonal group.

1 Implicit Function Theorem

Let M be an $m \times n$ real matrix. The so-called Gaussian elimination procedure provides a proof of the following fundamental fact:

There exist invertible matrices A and B such that

$$AMB = \begin{pmatrix} I_k & 0 \\ 0 & 0 \end{pmatrix}, \qquad k = \text{rank of } M.$$

If $L: \mathbf{R}^m \to \mathbf{R}^n$ is a linear map given by the matrix M, that is,

$$L(v) = M \cdot v,$$

then this result may be interpreted as saying that with respect to some system of coordinates in $\mathbf{R}^m$ and $\mathbf{R}^n$ the map L is the composition of the projection $\mathbf{R}^m \to \mathbf{R}^k$ with the inclusion $\mathbf{R}^k \hookrightarrow \mathbf{R}^n$. We will show that there is an appropriate generalization of this to arbitrary smooth maps $\mathbf{R}^m \to \mathbf{R}^n$.

Definition Let $f: U \to \mathbf{R}^m$ be a smooth map, U an open subset of $\mathbf{R}^m$, and let $p \in U$. If there is a neighborhood V of p such that the map $f \mid V: V \to f(V)$ has a smooth inverse, then we shall say that f is a *local diffeomorphism* at p. A local diffeomorphism at $\mathbf{0}$ is also called a *local coordinate system* at $f(\mathbf{0})$.

All that follows will be based on the following fundamental result:

(1.1) Implicit Function Theorem *Let $f: U \to \mathbf{R}^m$ be a smooth map, U an open subset of $\mathbf{R}^m$. If the rank of the Jacobian matrix $J(f, p)$ of f at $p \in U$ equals m, then f is a local diffeomorphism at p.*

Briefly: If $J(f, p)$ is invertible, then so also is f in a neighborhood of p.

The generalization of the linear algebra statement, as well as of the Implicit Function Theorem, that we are after is as follows:

(1.2) Theorem *Let $f: U \to \mathbf{R}^n$ be a smooth map, U an open subset of $\mathbf{R}^m$. If the Jacobian $J(f)$ is of constant rank k in a neighborhood of $p \in U$, then there is a local coordinate system g at p and a local coordinate system h at $f(p)$ such that*

$$h^{-1}fg(x_1, \ldots, x_m) = (x_1, \ldots, x_k, \mathbf{0}).$$

In other words, with respect to these coordinate systems the map f is the composition of the projection $\mathbf{R}^m \to \mathbf{R}^k$ with the inclusion $\mathbf{R}^k \hookrightarrow \mathbf{R}^n$.

Proof Without restricting the generality of the argument we can assume that $p = \mathbf{0} \in \mathbf{R}^m$ and $f(p) = \mathbf{0} \in \mathbf{R}^n$. Let $f = (f_1, \ldots, f_n)$. Renumbering the variables we can also achieve that the determinant of the matrix

$$J_1 = \begin{pmatrix} \dfrac{\partial f_1}{\partial x_1} & \cdots & \dfrac{\partial f_1}{\partial x_k} \\ & \cdots & \\ \dfrac{\partial f_k}{\partial x_1} & \cdots & \dfrac{\partial f_k}{\partial x_k} \end{pmatrix}$$

is not zero at $\mathbf{0}$.

Let $f' = (f_1, \ldots, f_k)$, $f'' = (f_{k+1}, \ldots, f_n)$ and set $F(x_1, \ldots, x_m) = (f', x_{k+1}, \ldots, x_m)$. Since

$$J(F, \mathbf{0}) = \begin{pmatrix} J_1 & * \\ 0 & I_{m-k} \end{pmatrix}$$

is non-singular, there is a local diffeomorphism g defined in a neighborhood of $\mathbf{0}$ such that $Fg(x) = x$, and hence $f'g(x) = (x_1, \ldots, x_k)$.

If $k = n$, then $f = f'$ and the proof ends here; otherwise we continue as follows. Since $fg(x) = (x_1, \ldots, x_k, f''g(x))$,

$$J(fg, \mathbf{0}) = \begin{pmatrix} I_k & 0 \\ * & J_2 \end{pmatrix},$$

where J_2 is the Jacobian of $f''g$ with respect to the variables $x_{k+1}, \ldots, x_m$. Since $J(fg) = J(f)J(g)$ and $J(g)$ is invertible, the rank of $J(fg)$ equals the rank of $J(f)$. The latter equals k in a neighborhood V of $\mathbf{0} \in \mathbf{R}^m$; thus J_2 has only zero entries in V. This means that, in V, $f''g$ is a function of the variables $x_1, \ldots, x_k$ only. Denote this function by g', let $\sigma(x_1, \ldots, x_n) = (x_1, \ldots, x_k)$, $\pi(x_1, \ldots, x_n) = (x_{k+1}, \ldots, x_n)$, and define

$$h(x_1, \ldots, x_n) = (\sigma(x), g'\sigma(x) - \pi(x)).$$

Then h is a local coordinate system at $\mathbf{0} \in \mathbf{R}^n$ and

$$fg(x) = h(\sigma(x), g'\sigma(x)) = h(x_1, \ldots, x_k, \mathbf{0}). \qquad \square$$

An often encountered case of 1.2 is when the Jacobian of f is of maximal rank at p, that is, of rank m or n. Since in this case the rank must be constant in a neighborhood of p, we have the following:

(1.3) Corollary *If f is of maximal rank at p, then with respect to appropriately chosen coordinate systems f is:*

—*the projection* $\mathbf{R}^m \to \mathbf{R}^n$, *if* $n \leq m$;
—*the inclusion* $\mathbf{R}^m \hookrightarrow \mathbf{R}^n$, *if* $m \leq n$. $\qquad \square$

All results of this section apply to smooth maps defined on an open subset of $\mathbf{R}_+^m = \{x \in \mathbf{R}^m \mid x_m \geq 0\}$, such a map being smooth if it extends—at least locally—to a smooth map on an open subset of $\mathbf{R}^m$. For if f is such a map, $p \in \mathbf{R}^{m-1}$ and $\bar{f}$ is an extension of f over a neighborhood of p in $\mathbf{R}^m$, then $J(\bar{f}, p)$ is completely determined by $\bar{f} \mid \mathbf{R}_+^m$, i.e., by f.

2 A Lemma of M. Morse

The investigation of maps $\mathbf{R}^m \to \mathbf{R}^n$ is frequently simplified by a lemma of M. Morse, which gives a presentation of such a map that looks like a linear map but has variable coefficients.

(2.1) Lemma *Let* $U \subset \mathbf{R}^m$ *be a convex neighborhood of* $\mathbf{0}$ *and* $f: U \to \mathbf{R}$ *a smooth function,* $f(\mathbf{0}) = \mathbf{0}$. *Then*

$$(*) \qquad\qquad f(x) = \sum_i a_i(x) x_i$$

for some smooth functions $a_1(x), \ldots, a_m(x)$ *satisfying* $a_i(\mathbf{0}) = \partial f / \partial x_i(\mathbf{0})$.

Proof $\quad f(x) = \int_0^1 \dfrac{d}{dt} f(tx)\, dt = \int_0^1 \sum_i \dfrac{\partial f(tx)}{\partial x_i} x_i\, dt = \sum_i \left(\int_0^1 \dfrac{\partial f(tx)}{\partial x_i}\, dt \right) x_i.$

The second part follows by differentiating $(*)$. $\qquad \square$

(2.2) Theorem *Let* $U \subset \mathbf{R}^m$ *be a convex neighborhood of* $\mathbf{0}$ *and* $f: U \to \mathbf{R}^n$ *a smooth map,* $f(\mathbf{0}) = \mathbf{0}$. *Then*

$$f(x) = M(x) \cdot x,$$

where $M(x)$ *is an* $m \times n$ *matrix whose entries are smooth functions of* x, *and* $M(\mathbf{0}) = J(f, \mathbf{0})$.

Proof Let $f = (f_1, \ldots, f_n)$ and apply 2.1 to each function f_i. $\qquad \square$

With $f(x)$ represented as in the preceding, consider the map

$$F_t(x) = M(tx) \cdot x, \qquad 0 \le t \le 1.$$

F_t is a deformation of the map f to the linear map given by the Jacobian of f at $\mathbf{0}$. Observe that the rank of $J(F_t, \mathbf{0})$ equals the rank of $J(f, \mathbf{0})$. This is so because $F_t(x) = (1/t) f(tx)$ for $t \ne 0$, i.e., F_t is the composition of f with the multiplication by t in $\mathbf{R}^m$ and the multiplication by $1/t$ in $\mathbf{R}^n$, and those two maps are of maximal rank.

3 Brown–Sard Theorem

We will derive a version of the Brown-Sard Theorem adapted to our purposes.

Recall that a subset C of $\mathbf{R}^m$ is said to be of measure 0 if, for every $\varepsilon > 0$, C is contained in a denumerable family of balls $\{B_j\}$ with total volume less than ε. A set of measure 0 cannot contain an open set; therefore its complement is dense in $\mathbf{R}^m$. Note also that if C_i is a denumerable family of sets of measure 0, then $C = \bigcup_i C_i$ is also of measure 0. For, if $\varepsilon > 0$ is given, then each set C_i is contained in a family $\{B_j^i\}$ of balls with total volume less than $\varepsilon / 2^i$. Then the family $\{B_j^i\}$ contains C and its total volume is less than ε.

Now, let $f : U \to \mathbf{R}^n$ be a smooth map, $U \subset \mathbf{R}^m$. A point $p \in U$ is a singular point of f if the rank of $J(f, p)$ is less than n. Let $S \subset U$ be the set of singular points of f. The theorem of Brown and Sard asserts that

$$f(S) \text{ is a measure } 0.$$

(The proof given by Pontriagin [Po2] presently enjoys great popularity. Simplified versions of it can be found in [M2], [H1], [Bd].)

This generalizes to maps of differentiable manifolds.

(3.1) Theorem *Let $f : M \to N$ be a smooth map of smooth manifolds. Then the set of regular values of f is dense in N.*

Recall that the set of regular values is the complement in N of $f(S)$, where S is the set of singular points of either f or $f \,|\, \partial M$, cf. II,2.4.

Proof Let $\{U_\alpha, h_\alpha\}$ be an adequate atlas on M and $\{V_\beta, g_\beta\}$ an adequate atlas on N. We have to show that, for every β, $V_\beta - f(S)$ is dense in V_β, i.e., that $g_\beta^{-1}(V_\beta - f(S))$ is dense in $g_\beta^{-1}(V_\beta)$. This will follow if we establish that $g_\beta^{-1}(V_\beta \cap f(S))$ is of measure 0. But $g_\beta^{-1}(V_\beta \cap f(S)) = \bigcup_\alpha g_\beta^{-1}(V_\beta \cap f(S \cap U_\alpha))$ and each of the sets $g_\beta^{-1}(V_\beta \cap f(S \cap U_\alpha))$ is of measure 0 by the Brown–Sard Theorem. $\square$

4 Orthonormalization

We give two theorems here about decompositions of matrices. The emphasis is on smoothness of procedure and uniqueness of results.

The following theorem is known as the *Gram–Schmidt orthonormalization procedure.*

(4.1) **Theorem** *Given a matrix $M \in \mathrm{Gl}(n)$, there exists a unique upper triangular matrix T with positive entries on the diagonal such that MT is an orthogonal matrix O. The entries of T are smooth functions of entries of M.*

Proof This is usually proved by induction; the geometric content of the inductive step can be described as follows. Suppose that the first k columns $v_1, \ldots, v_k$ are orthonormal and consider the vector space V_{k+1} spanned by $v_1, \ldots, v_k, v_{k+1}$. There is a unique vector w such that:
  (*) $v_1, \ldots, v_k, w$ form an orthonormal basis of V_{k+1};
  (**) The orientations of V_{k+1} given by $v_1, \ldots, v_k, v_{k+1}$ and $v_1, \ldots, v_k, w$
      coincide.
(w is obtained by subtracting from v_{k+1} all projections on $v_1, \ldots, v_k$ and normalizing the result; w depends smoothly on $v_1, \ldots, v_{k+1}$.)

Now, w is taken as the $(k + 1)$st column of O.

The uniqueness of O, hence of T, follows: The triangularity of T forces (*), and the fact that its diagonal elements are positive forces (**).   □

Let $\mathbf{T}(n)$ denote the set of upper triangular matrices with positive diagonal elements. Clearly, $\mathbf{T}(n)$ can be identified with a convex subset of $\mathbf{R}^{n(n+1)/2}$. Therefore we have:

(4.2) **Corollary** $\mathrm{Gl}(n) = \mathbf{O}(n) \times \mathbf{T}(n)$; $\mathbf{O}(n)$ *is a deformation retract of* $\mathrm{Gl}(n)$.   □

The decomposition $M = OT$ in 4.1 lacks an important property: If M and M_1 are orthogonally similar and $M = OT$, $M_1 = O_1 T_1$, then O and O_1 need not be orthogonally similar. The existence of a decomposition having this property is assured by the following theorem of Chevalley.

(4.3) **Theorem** *Given a matrix $M \in \mathrm{Gl}(n)$, there is a unique symmetric positive definite (s.p.d.) matrix S such that $M = OS$ with $O \in \mathbf{O}(n)$. S is a smooth function of M.*

Proof We first observe that $M = OS$ with O orthogonal and S symmetric if and only if there is a symmetric S such that $S^2 = {}^t MM$. For if $S^2 = {}^t MM$ with S symmetric, then letting $O = MS^{-1}$ we have ${}^t OO = {}^t(S^{-1}){}^t MSS^{-1} = I_n$. The converse is proved similarly.

Now, ${}^t MM$ is s.p.d. Hence to prove the theorem we have to show that an s.p.d. matrix has a unique s.p.d. square root depending smoothly on it.

Taking into account that the set $S(n)$ of s.p.d. matrices is an open subset of $\mathbf{R}^{n(n+1)/2}$, this amounts to asserting that the map $f : S(n) \to S(n)$ sending M to M^2 is a diffeomorphism. Thus we have to prove three statements:

(a) f is surjective;

(b) f is one-to-one (hence it has an inverse);

(c) f is of maximal rank at every point of $S(n)$ (hence, by 1.1, its inverse is smooth).

Let $O \in \mathbf{O}(n)$ and $f_O(M) = OMO^{-1}$. Then:

(4.3.1) $f_O: S(n) \to S(n)$ is a diffeomorphism and $f_O f = f f_O$.

This is clear since $f_{O^{-1}}$ is the inverse of f_O.

Now, if $M \in S(n)$, then, for some $O \in \mathbf{O}(n)$, $f_O(M)$ is a diagonal matrix $\operatorname{diag}(d_1, \ldots, d_n)$ with all d_i positive. Let $D = \operatorname{diag}(\sqrt{d_1}, \ldots, \sqrt{d_n})$. Then $f(f_O^{-1}D) = f_O^{-1}D^2 = f_O^{-1}f_O(M) = M$, which proves (a).

Now, let S be s.p.d. and assume that $S^2 = \operatorname{diag}(d_1, \ldots, d_n)$. To prove (b) we have to show that $S = \operatorname{diag}(\sqrt{d_1}, \ldots, \sqrt{d_n})$. Let $D = \operatorname{diag}(1/\sqrt{d_1}, \ldots, 1/\sqrt{d_n})$. We will show that $SD = I_n$ by showing that SD is symmetric with all eigenvalues $= +1$. The following argument was suggested to me by P. Landweber.

Observe first that $D^2S = SD^2$; hence $D(SD - DS) = (DS - SD)D$, i.e., $SD - DS$ anticommutes with a diagonal matrix with positive entries. This implies readily that $SD - DS = 0$; thus SD is a symmetric matrix. Since $(SD)^2 = I_n$, the eigenvalues of SD equal ± 1. But the eigenvalues of SD are $(e_1/\sqrt{d_1}, \ldots, e_n/\sqrt{d_n})$, where $(e_1, \ldots, e_n)$ are the eigenvalues of S. (This follows from the fact that an eigenspace of D is an invariant subspace for S.) Since all the e_i are positive, all eigenvalues of SD equal 1. This completes the proof of (b).

There remains to prove (c). By 4.3.1 it is enough to calculate the rank of the Jacobian $J(f, M)$ when M is a diagonal matrix.

We arrange the calculations as follows. Let $M = (x_{ij})$, $x_{ij} = x_{ji}$. The map f is given by $n(n + 1)/2$ functions

$$f_{11}, \ldots, f_{1n}, f_{22}, \ldots, f_{2n}, \ldots, f_{nn},$$

where f_{ij} is the product of the ith and jth rows of M. In $J(f, M)$ the derivatives of f_{ij} fill the $r(i, j)$th row, where

$$r(i, j) = n + (n - 1) + \cdots + (n - i + 2) + (j - i + 1),$$

and the derivative with respect to x_{mn} stands at the $r(m, n)$th place in this row. (Only x_{mn} with $n \geq m$ are considered.) This implies that entries below the diagonal are either 0 or x_{ij} with $i \neq j$ and the entries on the diagonal are the diagonal entries of M, with coefficient 2 if $i = j$. This implies that if M is a diagonal matrix, then $J(f, M)$ is upper triangular and its determinant equals 2^n times a product of diagonal entries of M. Thus, if M is s.p.d., $\det J(f, M) > 0$. This concludes the proof of (c) and of 4.3.   □

Another proof of 4.3 can be found in [Ch,1. § V]. But the preceding proof utilizes only elementary notions of linear algebra that can be found in most undergraduate texts.

(4.4)   Corollary $Gl(n) = O(n) \times S(n)$ *as topological spaces.*   □

Now let $0 \leq t \leq 1$ and let S_1, S_2 be two s.p.d. matrices. Then

$$\langle (tS_1 + (1 - t)S_2)v, v \rangle = t\langle S_1 v, v \rangle + (1 - t)\langle S_2 v, v \rangle > 0,$$

i.e., $S(n)$ is convex. An important consequence of this is given in I,3.3. (The fact that $O(n)$ is a deformation retract of $Gl(n)$ we already know from 4.2.)

5   Homotopy Groups of SO(k)

Let $\phi: SO(k) \to S^{k-1}$ associate to a matrix $M \in SO(k)$ its first column, i.e., $\phi(M) = M \cdot e_1$ where $e_1, \ldots, e_n$ is the standard basis of $\mathbf{R}^n$. Then ϕ is the bundle projection of a bundle with fiber $SO(k-1)$, with the inclusion $s_{k-1}: SO(k-1) \hookrightarrow SO(k)$ given by

$$M \mapsto \begin{pmatrix} 1 & 0 \\ 0 & M \end{pmatrix},$$

[S,7.6]. The homotopy exact sequence of this bundle implies that this inclusion induces an isomorphism $\pi_i(SO(k-1)) \to \pi_i(SO(k))$ for $i < k - 2$ and is surjective for $i = k - 2$. In particular, the groups $\pi_i(SO(k))$ stabilize for $k > i + 2$; we let $\pi_i(SO)$ denote their common value, i.e., $\lim \pi_i(SO(k))$. R. Bott showed in 1959 that it depends only on the congruence class of $i \bmod 8$ and calculated it as follows (*cf.* [M1] for a proof and references):

(5.1)

$i \bmod 8 =$	0	1	2	3	4	5	6	7
$\pi_i(SO) =$	$\mathbf{Z}_2$	$\mathbf{Z}_2$	0	$\mathbf{Z}$	0	0	0	$\mathbf{Z}$

Besides the stable group we will need information about the group $\pi_{k-1}(\mathbf{SO}(k))$. For this purpose we consider the diagram

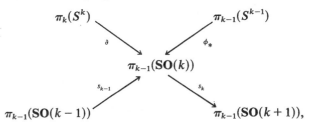

where the diagonal sequences are parts of homotopy exact sequences of the fibrations $\mathbf{SO}(k+1)/\mathbf{SO}(k) \to S^k$ and $\mathbf{SO}(k)/\mathbf{SO}(k-1) \to S^{k-1}$. Let ι_k be a generator of $\pi_k(S^k)$ and let $\partial \iota_k = \tau_k$; τ_k is the characteristic element of TS^k.

(5.2) Proposition (a) *If k is odd and $\neq 1, 3, 7$, then τ_k is of order 2; $\tau_k = 0$ if $k = 1, 3, 7$.*

(b) *If k is even, then $\phi_* \tau_k = 2\iota_{k-1}$; $\pi_{k-1}(\mathbf{SO}(k)) = \operatorname{Im} \partial \oplus \operatorname{Im} s_{k-1}$ unless $k = 2, 4, 8$.*

(c) *The composition $s_k s_{k-1}$ is surjective except if $k = 2, 4, 8$.*

Proof For (a) see [S,24.9]. The statement that S^k is not parallelizable for $k \neq 1, 3, 7$ can be found in [B3].

The first part of (b) is proved in [S,23.4]. To prove the second part observe first that $\pi_{k-1}(\mathbf{SO}(k)) \simeq \operatorname{Im} \phi_* \oplus \operatorname{Im} s_{k-1}$. To finish the proof it is enough to show that $\operatorname{Im} \phi_* = \operatorname{Im} \phi_* \partial$ for $k \neq 2, 4, 8$. This, in turn, is equivalent to proving that if ϕ_* is surjective, then $k = 2, 4, 8$. But if $\alpha \in \pi_{k-1}(\mathbf{SO}(k))$ is such that $\phi_*(\alpha) = \iota_{k-1}$, then $\tau_{k-1} = \partial \iota_{k-1} = \partial \phi_*(\alpha) = 0$; hence $k = 2, 4, 8$ by (a).

Now, for k odd $\pi_{k-1}(S^{k-1}) \to \pi_{k-2}(\mathbf{SO}(k-1))$ is injective by the first part of (b). Hence s_{k-1} is surjective, which proves (c) in this case. For k even (c) follows from the second part of (b), for $\operatorname{Ker} s_k = \operatorname{Im} \partial$. $\square$

If $k = 2, 4, 8$, then the situation is complicated by the presence of elements α with $\phi_*(\alpha) = \iota_{k-1}$, the Hopf fibrations. The argument used in the proof of (b) shows that $\pi_{k-1}(\mathbf{SO}(k)) = H(\alpha) \oplus \operatorname{Im} s_{k-1}$, where $H(\alpha)$ is generated by the α. If $k = 4$, then $\operatorname{Im} s_3$ is infinite cyclic with generator σ, and $\tau_4 = -\sigma + 2\alpha$, cf. [S,23.6]; hence the composition $s_4 s_3$ yields only the even elements of $\pi_3(\mathbf{SO}(5)) \simeq \mathbf{Z}$.

Appendix II

The Work of Grigory Perelman

John W. Morgan*

Introduction

I will report on the work of Grigory Perelman for which he was awarded the
Fields Medal at the International Congress in the summer of 2006. Perelman
posted three preprints on the arXiv between November 2002 and July 2003,
[14, 16, 15]. In these preprints he gave a complete, albeit highly condensed,
proof of the Poincaré Conjecture. Furthermore, at the end of the second
preprint he stated a theorem about three-manifolds with curvature bounded
below which are sufficiently collapsed. He showed how, from what he had
established in the first two of his preprints, this collapsing result would imply
the vast generalization of the Poincaré Conjecture, known as Thurston's Geometrization Conjecture. He stated that he would provide another manuscript
proving the collapsing result, and in private conversations, he indicated that
the proof used ideas contained in an earlier unpublished manuscript of his
from 1992 and a then recently circulated manuscript of Shioya-Yamaguchi
(which has now appeared [20]). To date, Perelman has not posted the follow-
up paper establishing the collapsing result he stated at the end of his second
preprint.

Copyright (c) 2007 by The American Mathematical Society. Originally published
in Notices of the AMS, Volume 54 (3), pp. 393-399, 2007. Reprinted by permission of
the American Mathematical Society.

*John W. Morgan is professor of mathematics at Columbia University. His email
address is jm@math.columbia.edu.

In this article, I will explain a little of the history and significance of the Poincaré Conjecture and Thurston's Geometrization Conjecture. Then I will describe briefly the methods Perelman used to establish these results, and I will discuss my view of the current state of the Geometrization Conjecture. Lastly, I will speculate on future directions that may arise out of Perelman's work. For a survey on the Poincaré Conjecture see [12]. For more details of the ideas and results we sketch here, the reader can consult [10], [2], and [13] and [23].

The Poincaré Conjecture and Thurston's Geometrization Conjecture

In 1904 Poincaré asked whether every closed (i.e., compact without boundary) simply connected three-manifold is homeomorphic to the three-sphere, see [18]. (It has long been known that this is equivalent to asking whether a simply connected smooth three-manifold is diffeomorphic to the three-sphere.) What has come to be known as the Poincaré Conjecture is the conjecture that the answer to this question is "yes". Since its posing, the Poincaré Conjecture has been a central problem in topology, and most of the advances in study of the topology of manifolds, both in dimensions three and in higher dimensions, over the last one hundred years have been related to the Poincaré Conjecture or its various generalizations. Prior to Perelman's work, analogues of the conjecture had been formulated in all dimensions and the topological versions of these analogues had been established (see [21] and [6]) in all dimensions except dimension three. Counterexamples to the analogue of the Poincaré Conjecture for smooth manifolds of higher dimension were first given by Milnor, in [11], where he constructed exotic smooth structures on the 7-dimensional sphere. The smooth four-dimensional Poincaré Conjecture remains open. Perelman's work resolves one remaining case of the topological version of the question. The Poincaré Conjecture is also the first of the seven Clay Millennium problems to be solved.

Poincaré was motivated to ask his question by an attempt to characterize the simplest of all three-manifolds, the three-sphere. But there is no reason to restrict only to this three-manifold. In 1982 Thurston formulated a general conjecture, known as Thurston's Geometrization Conjecture, which, if true, would (essentially) classify all closed three-manifolds, see [24] and [19]. Unlike the Poincaré Conjecture, which has a purely topological conclusion, Thurston's Conjecture has a geometric conclusion: In brief, it says that every closed three-manifold has an essentially unique two-step decomposition into simpler pieces. The first step is to cut the manifold open along a certain family of embedded two-spheres and cap off the resulting boundaries with three-balls. The second step is to cut the manifold open along a certain family of two-tori. The conjecture posits that at the end of this process each of the resulting pieces will admit a complete, finite-volume, Riemannian metric locally modeled on one of the eight homogeneous three-dimensional geometries. All manifolds modeled on seven of the eight geometries are completely understood and easily listed. The eighth homogeneous three-dimensional geometry is hyperbolic geometry. It is the richest class of examples and is the most interesting case. Manifolds modeled on this are given as quotients of hyperbolic three-space by torsion-free lattices in $SL(2, \mathbb{C})$ of finite co-volume. The classification of these manifolds is equivalent to the classification of such lattices, a problem that has not yet been completely solved. Thurston's Conjecture includes the Poincaré Conjecture as a special case, since the only geometry that can model a simply connected manifold is the spherical geometry (constant positive curvature), and simply connected manifolds of constant positive curvature are easily shown to be isometric to the three-sphere.

Over the last one hundred years there have been many attempts to prove the Poincaré Conjecture. Most have been direct topological attacks which involve simplifying surfaces and/or loops in a simply connected three-manifold in order to show that the manifold is the union of two three-balls, which then implies that it is homeomorphic to the three-sphere. While these types of topological arguments have proved many beautiful results about three-dimensional manifolds, including for example that a knot in the three-sphere is trivial (i.e., unknotted) if and only if the fundamental group of its complement is isomorphic to $\mathbb{Z}$, they have said nothing about the Poincaré Conjecture. For a description of some of these approaches and why they failed see [22]. More recently, there have been approaches to the Geometrization Conjecture along the following lines. Given any three-manifold, the complement of a sufficiently general knot in the manifold will have a complete hyperbolic structure of finite volume. Work from here back to the original manifold by navigating in the space of hyperbolic structures with cone-like singularities. This approach, pioneered by Thurston, proved a beautiful result, called the orbifold theorem [5], that proves a special case of the Geometrization Conjecture, but has not led to a proof of the full conjecture.

Perelman's Method: Ricci Flow

How did Perelman do it? He used, and generalized, the work of Richard Hamilton on Ricci flow. For an introduction to Ricci flow see [3] and for a collection of Hamilton's papers on Ricci flow see [1]. Hamilton showed, at least in good cases, that the Ricci flow gives an evolution of a Riemannian metric on a 3-manifold that converges to a locally homogeneous metric (e.g., a metric of constant positive curvature), see [7] and [9]. But Hamilton also showed (see [8]) that there are other possibilities for the effect of Ricci flow. It can happen, and does happen, that the Ricci flow develops finite-time singularities. These finite-time singularities impede the search for a good metric; for, in order to find the good metric one needs to continue the flow for all positive time—the Riemannian metrics one is looking for in general only appear as limits as time goes to infinity. In fact, the finite-time singularities are needed in order to do the first layer of cutting (along two-spheres and filling in balls) as required in Thurston's Conjecture. These cuttings along two-spheres and filling in balls have been long known to be necessary in order to produce a manifold that admits a Riemannian metric modeled on one of the homogeneous geometries. As we shall see presently, the second step in the decomposition proposed by Thurston, along two-tori, happens as t limits to infinity.

The Ricci flow equation, as introduced and first studied by Hamilton in [7], is a weakly parabolic partial differential equation for the evolution of a Riemannian metric g on a smooth manifold. It is given by:

$$\frac{\partial g(t)}{\partial t} = -2\mathrm{Ric}(g(t)),$$

where $\mathrm{Ric}(g(t))$ is the Ricci curvature of $g(t)$. One should view this equation as a (weakly) nonlinear version of the heat equation for symmetric two-tensors on a manifold. Hamilton laid down the basics of the theory of this equation—short-time existence and uniqueness of solutions. Furthermore, using a version of the maximum principle he established various important differential inequalities, including a Harnack-type inequality, that are crucial in Perelman's work. Hamilton also understood that, in dimension three, surgery would be a crucial ingredient for two reasons: (i) it is the way to deal with the finite-time singularities and (ii) because of the necessity in general to do cutting on a given three-manifold in order to find the good metric. With this motivation, Hamilton introduced the geometric surgeries that Perelman employs.

From the point of view of the previous paragraph, an n-dimensional Ricci flow is a one-parameter family of metrics on a smooth manifold. It is also fruitful to take the perspective of space-time and define an n-dimensional Ricci flow as a horizontal metric on $M \times [a, b]$ where the metric on $M \times \{t\}$ is the metric $g(t)$ on M. This point of view allows for a natural generalization: a generalized n-dimensional Ricci flow is an $(n + 1)$-dimensional space-time equipped with a time function which is a submersion onto an interval, a vector field which will allow us to differentiate in the "time direction", and a metric on the "horizontal distribution" (which is the kernel of the differential of the time function). We require that these data be locally isomorphic to a Ricci flow on a product $M \times (a, b)$, with the vector field becoming $\partial/\partial t$ in the local product structure.

Perelman's First Two Preprints

Here we give more details on Perelman's first two preprints, [14] and [16]. As we indicated above, the work in these preprints is built on the foundation of Ricci flow as developed by Hamilton. A central concept is the notion of an n-dimensional Ricci flow, or generalized Ricci flow, being κ-non-collapsed on scales $\leq r_0$ for some $\kappa > 0$. Let p be a point of space-time and let t be the time of p. Denote by $B(p, t, r)$ the metric ball of radius r in the t time-slice of space-time, and denote by $P(x, t, r, -r^2)$ the backwards parabolic neighborhood consisting of all flow lines backwards in time from time t to time $t - r^2$ starting at points of $B(x, t, r)$. This means that for any point p and any $r \leq r_0$ the following holds. If the norm of the Riemannian curvature tensor on $P(p, t, r, -r^2)$ is at most r^{-2} then the volume of $B(x, t, r)$ is at least κr^n. Here are the main new contributions of his first two preprints:

(1) He introduced an integral functional, called the reduced $\mathcal{L}$-length for paths in the space-time of a Ricci flow.

(2) Using (1) he proved that for every finite time interval I, a Ricci flow of compact manifolds parameterized by I are non-collapsed where both the measure of non-collapsing and the scale depend only on the interval and the geometry of the initial manifold.

(3) Using (1) and results of Hamilton's on singularity development, he classified, at least qualitatively, all models for singularity development for Ricci flows of compact three-manifolds at finite times. These models are the three-dimensional, ancient solutions (i.e., defined for $-\infty < t \leq 0$) of non-negative, bounded curvature that are non-collapsed on all scales ($r_0 = \infty$).

(4) He showed that any sequence of points (x_n, t_n) in the space-time of a three-dimensional Ricci flow whose times t_n are uniformly bounded above and such that the norms of the Riemannian curvature tensors $Rm(x_n, t_n)$ go to infinity (a so-called "finite-time blow-up sequence")

has a subsequence converging geometrically to one of the models from (3). This result gives neighborhoods, called "canonical neighborhoods", for points of sufficiently high scalar curvature. These neighborhoods are geometrically close to corresponding neighborhoods in non-collapsed ancient solutions as in (3). Their existence is a crucial ingredient necessary to establish the analytic and geometric results required to carry out and to repeat surgery.

(5) From the classification in (3) and the blow-up result in (4), he showed that surgery, as envisioned by Hamilton, was always possible for Ricci flows starting with a compact three-manifold.

(6) He extended all the previous results from the category of Ricci flows to a category of certain well-controlled Ricci flows with surgery.

(7) He showed that, starting with any compact three-manifold, repeatedly doing surgery and restarting the Ricci flow leads to a well-controlled Ricci flow with surgery defined for all positive time.

(8) He studied the geometric properties of the limits as t goes to infinity of the Ricci flows with surgery.

Let us examine these in more detail. Perelman's length function is a striking new idea, one that he has shown to be extremely powerful, for example, allowing him to prove non-collapsing results. Let $(M, g(t))$, $a \leq t \leq T$, be a Ricci flow. The reduced $\mathcal{L}$-length functional is defined on paths $\gamma \colon [0, \bar{\tau}] \to M \times [a, T]$ that are parameterized by backwards time, namely satisfying $\gamma(\tau) \in M \times \{T - \tau\}$ for all $\tau \in [0, \bar{\tau}]$. One considers

$$l(\gamma) = \frac{1}{2\sqrt{\bar{\tau}}} \int_0^{\bar{\tau}} \sqrt{\tau} \left(R(\gamma(\tau)) + |X_\gamma(\tau)|^2 \right) d\tau.$$

Here, $R(\gamma(\tau))$ is the scalar curvature of the metric $g(T - \tau)$ at the point in M that is the image under the projection into M of $\gamma(\tau)$, and $X_\gamma(\tau)$ is the projection into TM of the tangent vector to the path γ at $\gamma(\tau)$. The norm of $X_\gamma(\tau)$ is measured using $g(T - \tau)$. It is fruitful to view this functional as the analogue of the energy functional for paths in a Riemannian manifold. Indeed, there are $\mathcal{L}$-geodesics, $\mathcal{L}$-Jacobi fields, and a function $l_{(x,T)}(q, t)$ that measures the minimal reduced $\mathcal{L}$-length of any $\mathcal{L}$-geodesic from (x, T) to (q, t). Fixing (x, T), the function $l_{(x,T)}(q, t)$ is a Lipschitz function of (q, t) and is smooth almost everywhere.

Perelman proves an extremely important monotonicity result along $\mathcal{L}$-geodesics for a function related to this reduced length functional. Namely, suppose that W is an open subset in the time-slice $T - \bar{\tau}$, and each point of W is the endpoint of a unique minimizing $\mathcal{L}$-geodesic. Then he defined the reduced volume of W to be

$$\int_W (\bar{\tau})^{-n/2} e^{-l(x,T)(w,T-\bar{\tau})} dvol.$$

For each $\tau \in (0, \bar{\tau}]$ let $W(\tau)$ be the result of flowing W along the unique minimal geodesics to time $T - \tau$. Then the reduced volume of $W(\tau)$ is a monotone non-increasing function of τ. This is the basis of his proof of the non-collapsing result: Fix a point (x, T) satisfying the hypothesis of non-collapsing for some $r \leq r_0$. The monotonicity allows him to transfer lower bounds on reduced volume (from (x, T)) at times near the initial time (which is automatic from compactness) to lower bounds on the reduced volume times near T. From this, one proves the non-collapsing result at (x, T).

Perelman then turned to the classification of ancient three-dimensional solutions (ancient in the sense of being defined for all time $-\infty < t \leq 0$) of bounded, non-negative curvature that are non-collapsed on all scales. By geometric arguments he established that the space of based solutions of this type is compact, up to rescaling. Here the non-collapsed condition is crucial: After we rescale to make the scalar curvature one at the base point, this condition implies that the injectivity radius at the base point is bounded away from zero. There are several types of these ancient solutions. A fixed time section is of one of the following types—(i) a compact round three-sphere or a Riemannian manifold finitely covered by a round three-sphere, (ii) a cylinder which is a product of a round two-sphere with the line or a Riemannian manifold finitely covered by this product, (iii) a compact manifold diffeomorphic to S^3 or $\mathbb{R}P^3$ that contains a long neck which is approximately cylindrical, (iv) a compact manifold of bounded diameter and volume (when they are rescaled to have scalar curvature 1 at some point), and (v) a non-compact manifold of positive curvature which is a union of a cap of bounded geometry (modulo rescaling) and a cylindrical neck (approximately S^2 times a line) at infinity.

Using the non-collapsing result and delicate geometric limit arguments, Perelman showed that in a Ricci flow on compact three-manifolds any finite-time blow-up sequence has a subsequence which, after rescaling to make the scalar curvature equal to one at the base point, converges geometrically to a non-collapsed, ancient solution of bounded, non-negative curvature. (Geometric convergence means the following: Given a finite-time blow-up sequence (x_n, t_n), after replacing the sequence by a subsequence, there is a model solution with base point and scalar curvature at the base point equal to one such that for any $R < \infty$ the balls of radius R centered at the x_n in the nth rescaled flow converge smoothly to a ball of radius R centered at the base point in the given model.) For example, we could have a component of positive Ricci curvature. According to Hamilton's result [7] this manifold contracts to a point at the singular time and as it does so it approaches a round (i.e., constant positive curvature) metric. Thus, the model in this case is a round manifold with the same fundamental group. Perelman's result implies that there is a threshold so that all points of scalar curvature above the threshold have canonical neighborhoods modeled on corresponding neighborhoods in non-collapsed ancient solutions. This leads to geometric and analytic control in these neighborhoods, which in turn is crucial for the later arguments showing that surgery is always possible.

Now let us turn to surgery. An n-dimensional Ricci flow with surgery is a more general object. It consists of a space-time which has a time function. The level sets of the time function are called the time-slices and they are compact n-manifolds. This space-time contains an open dense subset that is a smooth manifold of dimension $n + 1$ equipped with a vector field and a horizontal metric that make this open dense set a generalized n-dimensional Ricci flow. At the singular times, the time-slices have points not included in the open subset which is a generalized Ricci flow. This allows the topology of the time-slices to change as the time evolves past a singular time. For example, in the case just described above, when a component is shrinking to a point, the effect of surgery is to remove entirely that component. A more delicate case is when the singularity is modeled on a manifold with a long, almost cylindrical tube in it. In this case, at the singular time, the singularities are developing inside the tube. Following Hamilton, at the singular time Perelman cuts off the tube near its ends and sews in a predetermined metric on the three-ball, using a partition of unity. The effect of surgery is to produce a new, usually topologically distinct manifold at the singular, or surgery, time. After having produced this new closed, smooth manifold, one restarts the Ricci flow using that manifold as the initial conditions. This then is the surgery process: remove some components of positive curvature and those fibered over circles by manifolds of positive curvature (all of which are topologically standard) and surger others along two-spheres, and then restart Ricci flow.

Perelman then showed the entire Ricci flow analysis described above extends to Ricci flows with surgery, provided that the surgery is done in a sufficiently controlled manner. Thus, one is able to repeat the argument ad infinitum and construct a Ricci flow with surgery defined for all time. Furthermore, one has fairly good control on both the change in the topology and the change in the geometry as one passes the surgery times. Here then is the main result of the first two preprints taken together:

Theorem *Let (M, g_0) be a compact, orientable Riemannian three-manifold. Then there is a Ricci flow with surgery $(\mathcal{M}, G)$ defined for all positive time whose zero-time slice is (M, g_0). This Ricci flow with surgery has only finitely many surgery times in any compact interval. As one passes a surgery time, the topology of the time-slices changes in the following manner. One does a finite number of surgeries along disjointly embedded two-spheres (removing an open collar neighborhood of the two-spheres and gluing three-balls along each of the resulting boundary two-spheres) and removes a finite number of topologically standard components (i.e., components admitting round metrics and components finitely covered by $S^2 \times S^1$).*

A couple of remarks are in order:

(i) It is clear from the construction that if Thurston's Geometrization Conjecture holds for the manifold at time t then it holds for the manifolds at all previous times.

(ii) In this theorem one does not need orientability, only the weaker condition that every projective plane has non-trivial normal bundle. To extend the results to cover manifolds admitting projective planes with trivial normal bundle one takes a double covering and works equivariantly. This fits perfectly with the formulation of the Geometrization Conjecture for such manifolds.

In addition, Perelman established strong geometric control over the nature of the metrics on the time-slices as time tends to infinity. In particular, at the end of the second preprint he showed that for t sufficiently large, the t time-slice contains a finite number of incompressible tori (incompressible means π_1 injective) that divide the manifold into pieces. Each component that results from the cutting process has a metric of one of two types. Either the metric is, after rescaling, converging to a complete constant negatively curved metric, or the metric is arbitrarily collapsed on the scale of its curvature. Components of the first type clearly support hyperbolic metrics of finite volume. It is to deal with the components of the second type that Perelman states the proposed result on collapsed manifolds with curvature bounded below.

Completion of the Proof of the Poincaré Conjecture

As we have already remarked, the Poincaré Conjecture follows as a special case of Thurston's Geometrization Conjecture. Thus, the results of the first two of Perelman's preprints together with the collapsing result stated at the end of the second preprint, give a proof of the Poincaré Conjecture. In a third preprint [15] Perelman gave a different argument, avoiding the collapsing result, proving the Poincaré Conjecture but not the entire Geometrization Conjecture. For a detailed proof along these lines, see Chapter 18 of [13]. One shows that if one begins with a homotopy three-sphere, or indeed any manifold whose fundamental group is a free product of finite groups and infinite cyclic groups, then the Ricci flow with surgery, which by the results of Perelman's first two preprints is defined for all positive time, becomes extinct after a finite time. That is to say, for all t sufficiently large, the manifold at time t is empty. We conclude that the original manifold is a connected sum of spherical space-forms (quotients of the round S^3 by finite groups of isometries acting freely) and S^2-sphere bundles over S^1. It follows that if the original manifold is simply connected, then it is diffeomorphic to a connected sum of three-spheres, and hence is itself diffeomorphic to the three-sphere. This shows that the Poincaré Conjecture follows from this finite-time extinction result together with the existence for all time of a Ricci flow with surgery.

Let us sketch how the finite-time extinction result is established in [15]. (There is a parallel approach in [4] using areas of harmonic two-spheres of non-minimal type instead of area minimizing 2-disks.) We consider the case of a homotopy three-sphere M (though the same ideas easily generalize to cover all cases stated above). The fact that the manifold is a homotopy three-sphere implies that $\pi_3(M)$ is non-trivial. Perelman's argument is to consider a non-trivial element in $\xi \in \pi_3(M)$. Represent this element by a two-sphere family of homotopically trivial loops. Then for each loop take the infimum of the areas of spanning disks for the loop, maximize over the loops in family and then minimize over all representative families for ξ. The result is an invariant $W(\xi)$. One asks what happens to this invariant under Ricci flow. The result (following similar arguments that go back to Hamilton [9]) is that

$$\frac{dW(\xi)}{dt} \leq -2\pi - \frac{1}{2} R_{\min}(t) W(\xi).$$

Here $R_{\min}(t)$ is the minimum of the scalar curvature at time t. Since one of Hamilton's results using the maximum principle is that $R_{\min}(t) \geq -6/(4t + \alpha)$ for some positive constant α, it is easy to see that any function satisfying Equation (1) goes negative in finite time. Perelman shows that the same equation holds in Ricci flows with surgery as long as the manifold continues to exist in the Ricci flow with surgery. On the other hand, $W(\xi)$ is always non-negative. It follows that the homotopy three-sphere must disappear in finite time.

This then completes Perelman's proof of the Poincaré Conjecture. Start with a homotopy three-sphere. Run the Ricci flow with surgery until the manifold disappears. Hence, it is a connected sum of manifolds admitting constant positive curvature metrics. That is to say it is a connected sum of three-spheres, and hence itself diffeomorphic to the three-sphere.

Status of the Geometrization Conjecture

What about the general geometrization theorem? Perelman's preprints show that proving this result has been reduced to proving a statement about manifolds with curvature locally bounded below that are collapsed in the sense that they have short loops through every point. There are results along these lines by Shioya-Yamaguchi [20]. The full result that Perelman needs can be found in an appendix to [20], except for the issue of allowing a boundary. The theorem that Perelman states in his second preprint allows for boundary tori, whereas the statement in [20] is for closed manifolds. Perelman has indicated privately that the arguments easily extend to cover this more general case. But in any event, invoking deep results from three-manifold topology, it suffices to consider only the closed case in order to derive the full Geometrization Conjecture. The paper [20] relies on an earlier, unpublished work by Perelman.

All in all, Perelman's collapsing result, in the full generality that he stated it, seems eminently plausible. Still, to my mind the entire collapsing space theory has not yet received the same careful scrutiny that Perelman's preprints have received. So, while I see no serious issues looming, I personally am not ready to say that geometrization has been completely checked in detail. I am confident that it is only a matter of time before these issues are satisfactorily explored.

The Effect of Perelman's Work

Let me say a few words about the larger significance of what Perelman has accomplished and what the future may hold. First of all, an affirmative resolution of the one-hundred-year-old Poincaré Conjecture is an accomplishment rarely equaled or surpassed in mathematics. An affirmative resolution of the Geometrization Conjecture will surely lead to a complete and reasonably effective classification of all closed three-dimensional manifolds. It is hard to overestimate the progress that this represents. The Geometrization Conjecture is a goal in its own right, leading as it (essentially) does to a classification of three-manifolds. Paradoxically, the effect of Perelman's work on three-manifold topology will be minimal. Almost all workers in the field were already assuming that the Geometrization Conjecture is true and were working modulo that assumption, or else they were working directly on hyperbolic three-manifolds, which obviously satisfy the Geometrization Conjecture.

The largest effects of Perelman's work will lie in other applications of his results and methods. I think there are possibilities for applying Ricci flow to four-manifolds. Four-manifolds are terra incognita compared to three-manifolds. In dimension four there is not even a guess as to what the possibilities are. Much less is known, and what is known suggests a far more complicated landscape in dimension four than in dimension three. In order to apply Ricci flow and Perelman's techniques to four-manifolds there are many hurdles to overcome. Nevertheless, this is an area that, in my view, shows promise.

There are also applications of Ricci flow to Kähler manifolds, where already Perelman's results are having an effect—see for example [17] and the references therein.

As the above description should make clear, Perelman's great advance has been in finding a way to control and qualitatively classify the singularities that develop in the Ricci flow evolution equation. There are many evolution equations in mathematics and in the study of physical phenomena that are of the same general nature as the Ricci flow equation. Some, such as the mean curvature flow, are related quite closely to the Ricci flow. Singularity development is an important aspect of the study of almost all of these equations both in mathematics and in physical applications. It is not yet understood if the type of analysis that Perelman carried out for the Ricci flow has analogues in some of these other contexts, but if there are analogues, the effect of these on the study of those equations could be quite remarkable.

References

[1] H. D. CAO, B. CHOW, S. C. CHU, and S. T. YAU, eds., *Collected papers on Ricci flow*, volume 37 of Series in Geometry and Topology, International Press, Somerville, MA, 2003.

[2] HUAI-DONG CAO and XI-PING ZHU, A complete proof of the Poincaré and Geometrization conjectures—Application of the Hamilton-Perleman theory of the Ricci flow, *Asian J. of Math* (2006), 169–492.

THE WORK OF GRIGORY PERELMAN

[3] BENNETT CHOW and DAN KNOPF, *The Ricci flow: an introduction*, volume 110 of Mathematical Surveys and Monographs, American Mathematical Society, Providence, RI, 2004.

[4] TOBIAS H. COLDING and WILLIAM P. MINICOZZI II, Estimates for the extinction time for the Ricci flow on certain 3-manifolds and a question of Perelman, *J. Amer. Math. Soc.* **18**(3) (2005), 561–569 (electronic).

[5] DARYL COOPER, CRAIG D. HODGSON, and STEVEN P. KERCKHOFF, *Three-dimensional orbifolds and cone-manifolds*, volume 5 of MSJ Memoirs, Mathematical Society of Japan, Tokyo, 2000, with a postface by Sadayoshi Kojima.

[6] MICHAEL HARTLEY FREEDMAN, The topology of four-dimensional manifolds, *J. Differential Geom.* **17**(3) (1982), 357–453.

[7] RICHARD S. HAMILTON, Three-manifolds with positive Ricci curvature, *J. Differential Geom.* **17**(2) (1982), 255–306.

[8] ————, The formation of singularities in the Ricci flow, in *Surveys in differential geometry, Vol. II (Cambridge, MA, 1993)*, Internat. Press, Cambridge, MA, 1995, pp. 1–136.

[9] ————, Non-singular solutions of the Ricci flow on three-manifolds, *Comm. Anal. Geom.* **7**(4) (1999), 695–729.

[10] BRUCE KLEINER and JOHN LOTT, Notes on Perelman's papers, arXiv:math.DG/0605667, 2006.

[11] JOHN MILNOR, On manifolds homeomorphic to the 7-sphere, *Ann. of Math. (2)* (1956), 399–405.

[12] ————, Towards the Poincaré conjecture and the classification of 3-manifolds, *Notices Amer. Math. Soc.* **50**(10) (2003), 1226–1233.

[13] JOHN MORGAN and GANG TIAN, Ricci flow and the Poincaré Conjecture, arXiv:math.DG/0607607, 2006.

[14] GRISHA PERELMAN, The entropy formula for the Ricci flow and its geometric applications, arXiv:math.DG/0211159, 2002.

[15] ————, Finite extinction time for the solutions to the Ricci flow on certain three-manifolds, arXiv:math.DG/0307245, 2003.

[16] ————, Ricci flow with surgery on three-manifolds, arXiv:math.DG/0303109, 2003.

[17] DUONG H. PHONG and JACOB STURM, On stability and the convergence of the Kähler-Ricci flow, *J. Differential Geom.* **72**(1) (2006), 149–168.

[18] HENRI POINCARÉ, Cinquième complément à l'analysis situs, In *Œuvres. Tome VI*, Les Grands Classiques Gauthier-Villars, [Gauthier-Villars Great Classics], pages v+541, Éditions Jacques Gabay, Sceaux, 1996, reprint of the 1953 edition.

[19] PETER SCOTT, The geometries of 3-manifolds, *Bull. London Math. Soc.* **15**(5) (1983), 401–487.

[20] TAKASHI SHIOYA and TAKAO YAMAGUCHI, Volume collapsed three-manifolds with a lower curvature bound, *Math. Ann.* **333**(1) (2005), 131–155.

[21] STEPHEN SMALE, Generalized Poincaré's conjecture in dimensions greater than four, *Ann. of Math.* **74**(2) (1961), 391–406.

[22] JOHN STALLINGS, How not to prove the Poincaré Conjecture, In *Topology Seminar of Wisconsin*, volume 60 of *Annals of Math Studies*,1965.

[23] TERRY TAO, Perelman's proof of the Poincaré Conjecture—a nonlinear PDE perspective, arXiv:math.DG/0610903, 2006.

[24] WILLIAM P. THURSTON, Hyperbolic structures on 3-manifolds. I, Deformation of acylindrical manifolds, *Ann. of Math.* **124**(2) (1986), 203–246.

Bibliography

[Ad]   J. F. Adams, On the groups $J(X)-IV$. *Topology* **5** (1966), 21-71.
[A]   J. W. Alexander, Some Problems in Topology. Verhandlungen Int. Math. Kongr., Zürich, 1932.
[AH]   P. Alexandroff and H. Hopf, Topologie. Springer, Berlin, 1935.
[B1]   R. Bott, Lectures on Characteristic Classes and Foliations. Lecture Notes in Math. 279. Springer, Berlin-Heidelberg-New York, 1972.
[B2]   R. Bott, Marston Morse and his mathematical works. *Bull. Amer. Math. Soc.* **3** (1980), 907-950.
[B3]   R. Bott and J. Milnor, On the parallelizability of the spheres. *Bull. Amer. Math. Soc.* **64** (1958), 87-89.
[Bd]   G. Bredon. Topology and Geometry. Springer, New York, 1993.
[Bk]   E. Brieskorn. Beispiele zur Differential topologie von Singularitäten. *Invent. Math.* **2** (1966), 1-14.
[Bo]   W. M. Boothby, An Introduction to Differentiable Manifolds and Riemannian Geometry. Academic Press, New York, 1975.
[Br1]   W. Browder, Surgery on Simply-Connected Manifolds. Springer, New York-Heidelberg-Berlin, 1972.
[Br2]   W. Browder, The Kervaire invariant of framed manifolds and its generalization. *Ann. of Math.* **90** (1969), 157-186.
[Br3]   W. Browder, On the action of $\theta^n(\pi)$. Differential and Combinatorial Topology, 23-36. Princeton University Press, Princeton, 1965.

[Bw]   M. Brown, A proof of the generalized Schoenflies theorem. *Bull. Amer. Math. Soc.* **66** (1960), 74–76.

[Ca]   S. S. Cairns, On the triangulation of regular loci. *Ann. of Math.* **35** (1934), 579–587.

[C1]   J. Cerf, Topologie des certains espaces de plongements. *Bull. Soc. Math. France*, **89** (1961), 227–380.

[C2]   J. Cerf, Sur les difféomorphismes de la sphère de dimension trois. Lecture Notes in Math. 53. Springer, Berlin–Heidelberg–New York, 1968.

[Ch]   C. Chevalley, Theory of Lie Groups I. Princeton University Press, Princeton, New Jersey, 1946.

[Cho]   G. Chogoshvili, Behaviour of some topological invariants on level surfaces. *Soob. Akad. Nauk Gruzinskoi SSR* **3** (1942), 995–999.

[Co]   M. Cohen, A Course in Simple-Homotopy Theory. Springer, New York–Heidelberg–Berlin, 1973.

[Di]   J. Dieudonné, A History of Algebraic and Differential Topology. Birkhäuser, Boston, Massachusetts, 1989.

[D]   A. Dold, Lectures on Algebraic Topology. Springer, New York–Heidelberg–Berlin, 1972.

[DK]   S. K. Donaldson and P. B. Kronheimer, The geometry of four-manifolds. Clarendon Press, Oxford, 1990.

[Du]   J. Dugundji, Topology. Allyn and Bacon, Boston, 1966.

[E1]   L. Euler, Introductio in Analysin Infinitorum. Bousquet, Lausanne, 1748.

[E2]   L. Euler, De solidis quorum superficiem in planum explicare licet. Opera Omnia I 28. Orel Fussli, Zurich, 1955.

[ES]   S. Eilenberg and N. Steenrod, Foundations of Algebraic Topology. Princeton University Press, Princeton, New Jersey, 1952.

[Eh]   C. Ehresmann, Les connexions infinitésimales dans un espace fibré différentiable. Colloque de Topologie, Bruxelles 1950, 29–55.

[F]   M. Freedman, The topology of four-dimensional manifolds. *J. Diff. Geom.* **17** (1982), 367–453.

[FQ]   M. Freedman and F. Quinn, The topology of 4-manifolds. Princeton University Press, Princeton, New Jersey, 1990.

[G]   C. F. Gauss, Disquisitiones generales circa superficies curvas. Werke 4, Göttingen, 1880.

[Go]   C. Godbillon, Feuilletages: Études Géométriques. Birkhauser, Basel-Boston, 1991.

[H1]   A. Haefliger, Differentiable imbeddings. *Bull. Amer. Math. Soc.* **67** (1961), 109–112.

[H2]   A. Haefliger, Knotted $(4k-1)$-spheres in $6k$-space. *Ann. of Math.* **75** (1962), 452–466.

[Hi]   M. Hirsch, Differential Topology. Springer, New York-Heidelberg-Berlin, 1976.

[Hr1]   F. Hirzebruch, Topological Methods in Algebraic Geometry. Springer, New York, 1966.

[Hr]   F. Hirzebruch und K. M. Mayer, $O(n)$-Mannigfaltigkeiten, exotische Sphären und Singularitäten. Lecture Notes in Math. 57. Springer, Berlin-Heidelberg-New York, 1968.

[Hl]   P. Holm, The Theorem of Brown and Sard. *Enseignement Math.* 33 (1987), 199-202.

[Ho1]   H. Hopf, Zur Topologie der Abbildungen von Mannigfaltigkeiten, *Math. Ann.* 102 (1929), 562-623.

[Ho2]   H. Hopf, Uber die Abbildungen der dreidimensionalen Sphäre auf die Kugelfläche. *Math. Ann.* 104 (1931), 637-665.

[Hu]   W. Hurewicz, Lectures on Ordinary Differential Equations. MIT Press, Cambridge, Massachusetts, 1958.

[K1]   M. Kervaire, An interpretation of G. Whitehead's generalization of H. Hopf's invariant. *Ann. of Math.* 69 (1959), 345-365.

[K2]   M. Kervaire, A manifold which does not admit any differentiable structure. *Comment. Math. Helv.* 34 (1960), 257-270.

[K3]   M. Kervaire, Le théorème de Barden-Mazur-Stallings. *Comment. Math. Helv.* 40 (1965), 31-42.

[KM1]   M. Kervaire and J. Milnor, On 2-spheres in 4-manifolds, *Proc. Nat. Acad. Sci. USA* 47 (1961), 1651-1657.

[KM2]   M. Kervaire and J. Milnor, Groups of homotopy spheres. *Ann. of Math.* 77 (1963), 504-537.

[Ki]   R. Kirby, The topology of 4-manifolds. Lecture Notes in Math. 1374. Springer, New York, 1989.

[KS]   R. Kirby and L. Siebenmann, Foundational essays on topological manifolds, smoothings, and triangulations. *Ann. of Math. Studies* 88. Princeton University Press, Princeton, New Jersey, 1977.

[Ko]   A. Kosinski, On the inertia group of π-manifolds. *Amer. J. of Math.* 89 (1967), 227-248.

[L]   S. Lang, Differential Manifolds. Springer, New York-Berlin-Heidelberg, 1985.

[Le1]   J. P. Levine, A classification of differentiable knots. *Ann. of Math.* 82 (1965), 15-50.

[Le2]   J. P. Levine, Lectures on groups of homotopy spheres. Algebraic and Geometric Topology Proceedings, Rutgers 1983. Lecture Notes in Math. 1126. Springer, Berlin-Heidelberg-New York, 1985.

[Ma1]   B. Mazur, On embedings of spheres. *Bull. Amer. Math. Soc.* 65 (1959), 59-65.

[Ma2]   B. Mazur, Simple neighborhoods. *Bull. Amer. Math. Soc.* **68** (1962), 87–92.

[Ma3]   B. Mazur, A note on some contractible 4-manifolds. *Ann. of Math.* **73** (1961), 221–228.

[M1]   J. Milnor, Morse theory. *Ann. of Math. Studies* **51**. Princeton University Press, Princeton, New Jersey, 1963.

[M2]   J. Milnor, Topology from the Differentiable Viewpoint. The University Press of Virginia, Charlottesville, Virginia, 1965.

[M3]   J. Milnor, On manifolds homeomorphic to the 7-sphere. *Ann. of Math.* **64** (1956), 399–405.

[M4]   J. Milnor, Differentiable structures on spheres. *Amer. J. of Math.* **81** (1959), 962–972.

[M5]   J. Milnor, Sommes de variétes différentiables et structures différentiables des sphères. Bull. Soc. Math. France **87** (1959), 439–444.

[M6]   J. Milnor, Two complexes which are homeomorphic but combinatorially distinct. *Ann. of Math.* **74** (1961), 575–590.

[M7]   J. Milnor, A procedure for killing homotopy groups of differentiable manifolds. *Proc. Sympos. Pure Math.* **3**, Amer. Math. Soc., Providence, Rhode Island, 1963.

[M8]   J. Milnor, Lectures on the H-Cobordism Theorem. Princeton University Press, Princeton, New Jersey, 1965.

[M9]   J. Milnor, Singular points of complex hypersurfaces. *Annals of Math. Studies* **61**. Princeton University Press, Princeton, New Jersey, 1968.

[MK]   J. Milnor and M. Kervaire, Bernoulli numbers, homotopy groups and a theorem of Rohlin. Proc. ICM 1962, Edinburgh, 1958.

[MS]   J. Milnor and J. Stasheff, Characteristic classes. *Ann. of Math. Studies* **76**. Princeton University Press, Princeton, New Jersey, 1974.

[Mo1]   M. Morse, Relations between the critical points of a real function of n independent variables. *Trans. Amer. Math. Soc.* **27** (1925), 345–396.

[Mo2]   M. Morse, Calculus of variations in the large. *Amer. Math. Soc. Colloq. Publ.* 18., Providence, Rhode Island, 1934.

[Mo3]   M. Morse, The existence of polar non-degenerate functions on differentiable manifolds. *Ann. of Math.* **71** (1960), 352–383.

[Mö]   A. F. Möbius. Theorie der elementaren Verwandschaften. Werke 3, Hirzel, Leipzig, 1886.

[Mu1]   J. Munkres, Differentiable isotopies on the 2-sphere. *Mich. Math. J.* **7** (1960), 193–197.

[Mu2]   J. Munkres, Obstructions to the smoothing of piecewise-differentiable homeomorphisms. *Ann. of Math.* **72** (1960), 521–554.

[N]   M. Newman, Integral Matrices. Academic Press, New York and London, 1972.

[Pa1]   R. Palais, Extending diffeomorphisms. *Proc. Amer. Math. Soc.* **11** (1960), 274–277.

[Pa2]   R. Palais, Local triviality of the restriction map for embeddings. *Comment. Math. Helv.* **34** (1960), 305–312.

[PR]   T. Petrie and J. D. Randall, Transformation Groups on Manifolds. Marcel Dekker, New York and Basel, 1984.

[P1]   H. Poincaré, Analysis Situs. *J. Ecole Polytechn.* **1** (1895), 1–121. (=Oeuvres 6, Gauthier-Villars, Paris, 1953.)

[P2]   H. Poincaré, Complément à l'Analysis Situs. *Rend. Circ. Mat. Palermo* **13** (1899), 285–343. (=Oeuvres 6, Gauthier-Villars, Paris, 1953.)

[P3]   H. Poincaré, Cinquième complément à l'Analysis Situs. *Rend. Circ. Mat. Palermo* **18** (1904), 45–110. (=Oeuvres 6, Gauthier-Villars, Paris, 1953.)

[Pi]   E. Pitcher, Inequalities of critical point theory. *Bull. Amer. Math. Soc.* **64** (1958), 1–30.

[Pn]   J. Pont, La Topologie Algébrique des Origines à Poincaré. Presses Universitaires de France, Paris, 1974.

[Po1]   L. S. Pontriagin, Classification of continuous maps of a complex into a sphere. *Dokl. Akad. Nauk SSSR* **19** (1938), 361–363.

[Po2]   L. S. Pontriagin, Smooth manifolds and their applications in homotopy theory. *Trudy Mat. Inst. Steklov* **45** (1955). (=*Amer. Math. Soc. Transl.* (2) **11**, 1955.)

[R]   B. Riemann, Uber die Hypothesen, welche der Geometrie zu Grunde liegen. Werke, Teubner, Leipzig, 1892.

[S]   N. Steenrod, The Topology of Fibre Bundles. Princeton University Press, Princeton, New Jersey, 1951.

[SE]   N. Steenrod and D. B. A. Epstein, Cohomology Operations. Princeton University Press, Princeton, New Jersey, 1962.

[ST]   H. Seifert and W. Threlfall, Lehrbuch der Topologie. Teubner, Leipzig, 1934.

[Se]   J.-P. Serre, Cours d'Arithmétique. Presses Universitaires de France, Paris, 1970.

[Si]   L. Siebenmann, Topological manifolds. *Proc. ICM 1970*, Gauthier-Villars, Paris, 1971.

[Sh]   V. V. Sharko, Functions on Manifolds. Naukova Dumka, Kiev, 1990.

[Sm1]   S. Smale, Diffeomorphisms of the 2-sphere. *Proc. Amer. Math. Soc.* **10** (1959), 621–626.

[Sm2]   S, Smale, The generalized Poincaré conjecture in higher dimensions. *Bull. Amer. Math. Soc.* **66** (1960), 373–375.

[Sm3]   S. Smale, On the structure of manifolds. *Amer. J. Math.* **84** (1962), 387–399.

[Sm4]   S. Smale, On the structure of 5-manifolds. *Ann. of Math.* **75** (1962), 38–46.

[Sm5]   S. Smale, A survey of some recent developments in differential topology.
            Bull. Amer. Math. Soc. **69** (1963), 131–145.
[Sm6]   S. Smale, The story of the higher dimensional Poincaré conjecture.
            Math. Intelligencer **12** (1990), 44–51.
[Sp]   E. Spanier, Algebraic Topology. McGraw-Hill, New York, 1966.
[St]   J. Stallings, Polyhedral homotopy spheres. *Bull. Amer. Math. Soc.* **66**
            (1960), 485–488.
[So]   R. Stong, Notes on Cobordism Theory. Princeton University Press,
            Princeton, New Jersey, 1968.
[Su]   W. A. Sutherland, A note on the parallelizability of sphere-bundles over
            spheres. *J. London Math. Soc.* **39** (1964), 55–62.
[T1]   R. Thom, Sur une partition en cellules associée a une fonction. *C. R.
            Acad. Sci. Paris* **228** (1949), 973–975.
[T2]   R. Thom, Quelques propriétés globales des variétés différentiables.
            Comment. Math. Helv. **29** (1954), 17–85.
[T3]   R. Thom, Les singularités des applications différentiables. *Ann. Inst.
            Fourier, Grenoble* **6** (1955–1956), 43–87.
[T4]   R. Thom, La classification des immersions. Sémin. Bourbaki **157**, 1957–
            58.
[T5]   R. Thom, Les classes characteristiques des variétés triangulées. Symp.
            Intern. de Topologia, 54–67. Mexico City, 1958.
[To]   H. Toda, Composition methods in homotopy groups of spheres. *Ann.
            of Math. Studies* **49**. Princeton University Press, Princeton, New Jersey,
            1962.
[VW]   O. Veblen and J. H. C. Whitehead, The Foundations of Differential
            Geometry. Cambridge Univ. Press, 1932.
[W1]   C. T. C. Wall, Killing the middle homotopy groups of odd dimensional
            manifolds. *Trans. Amer. Math. Soc.* **103** (1962), 421–433.
[W2]   C. T. C. Wall, Classification of $(n-1)$-connected $2n$-manifolds. *Ann.
            of Math.* **75** (1962), 163–189.
[W3]   C. T. C. Wall, Surgery on Compact Manifolds. Academic Press, London
            and New York, 1970.
[Wa]   A. H. Wallace, Modifications and cobounding manifolds. *Canad. J.
            Math.* **12** (1960), 503–528.
[Wh1]   J. H. C. Whitehead, Simplicial spaces, nuclei and m-groups. *Proc.
            London Math. Soc.* **45** (1939), 243–327.
[Wh2]   J. H. C. Whitehead, On the homotopy type of manifolds. *Ann. of Math.*
            41 (1940), 825–832.
[Wh3]   J. H. C. Whitehead, On C^1-complexes. *Ann. of Math.* **41** (1940), 809–824.
[Wh4]   J. H. C. Whithead, On the homotopy type of ANR's. *Bull. Amer. Math.
            Soc.* **54** (1948), 1133–1145.

[Wi1]   H. Whitney, Sphere spaces. *Proc. Nat. Acad. of Sci. USA* **21** (1935), 464–468.

[Wi2]   H. Whitney, Differentiable manifolds. *Ann. of Math.* **37** (1936), 645–680.

[Wi3]   H. Whitney, The self-intersections of a smooth manifold in $2n$ space. *Ann. of Math.* **45** (1949), 220–246.

[Wi4]   H. Whitney, Singularities of a smooth n-manifold in $(2n-1)$-space. *Ann. of Math.* **45** (1949), 247–293.

[Wi5]   H. Whitney, On singularities of mappings of Euclidean spaces. *Ann. of Math.* **62** (1955), 374–410.

Index

A CATALOG OF SELECTED
DOVER BOOKS
IN SCIENCE AND MATHEMATICS

Mathematics

FUNCTIONAL ANALYSIS (Second Corrected Edition), George Bachman and Lawrence Narici. Excellent treatment of subject geared toward students with background in linear algebra, advanced calculus, physics and engineering. Text covers introduction to inner-product spaces, normed, metric spaces, and topological spaces; complete orthonormal sets, the Hahn-Banach Theorem and its consequences, and many other related subjects. 1966 ed. 544pp. 6⅛ x 9¼. 0-486-40251-7

ASYMPTOTIC EXPANSIONS OF INTEGRALS, Norman Bleistein & Richard A. Handelsman. Best introduction to important field with applications in a variety of scientific disciplines. New preface. Problems. Diagrams. Tables. Bibliography. Index. 448pp. 5⅜ x 8½. 0-486-65082-0

VECTOR AND TENSOR ANALYSIS WITH APPLICATIONS, A. I. Borisenko and I. E. Tarapov. Concise introduction. Worked-out problems, solutions, exercises. 257pp. 5⅜ x 8¼. 0-486-63833-2

AN INTRODUCTION TO ORDINARY DIFFERENTIAL EQUATIONS, Earl A. Coddington. A thorough and systematic first course in elementary differential equations for undergraduates in mathematics and science, with many exercises and problems (with answers). Index. 304pp. 5⅜ x 8½. 0-486-65942-9

FOURIER SERIES AND ORTHOGONAL FUNCTIONS, Harry F. Davis. An incisive text combining theory and practical example to introduce Fourier series, orthogonal functions and applications of the Fourier method to boundary-value problems. 570 exercises. Answers and notes. 416pp. 5⅜ x 8½. 0-486-65973-9

COMPUTABILITY AND UNSOLVABILITY, Martin Davis. Classic graduate-level introduction to theory of computability, usually referred to as theory of recurrent functions. New preface and appendix. 288pp. 5⅜ x 8½. 0-486-61471-9

ASYMPTOTIC METHODS IN ANALYSIS, N. G. de Bruijn. An inexpensive, comprehensive guide to asymptotic methods—the pioneering work that teaches by explaining worked examples in detail. Index. 224pp. 5⅜ x 8½ 0-486-64221-6

APPLIED COMPLEX VARIABLES, John W. Dettman. Step-by-step coverage of fundamentals of analytic function theory—plus lucid exposition of five important applications: Potential Theory; Ordinary Differential Equations; Fourier Transforms; Laplace Transforms; Asymptotic Expansions. 66 figures. Exercises at chapter ends. 512pp. 5⅜ x 8½. 0-486-64670-X

INTRODUCTION TO LINEAR ALGEBRA AND DIFFERENTIAL EQUATIONS, John W. Dettman. Excellent text covers complex numbers, determinants, orthonormal bases, Laplace transforms, much more. Exercises with solutions. Undergraduate level. 416pp. 5⅜ x 8½. 0-486-65191-6

RIEMANN'S ZETA FUNCTION, H. M. Edwards. Superb, high-level study of landmark 1859 publication entitled "On the Number of Primes Less Than a Given Magnitude" traces developments in mathematical theory that it inspired. xiv+315pp. 5⅜ x 8½. 0-486-41740-9

INTRODUCTORY REAL ANALYSIS, A.N. Kolmogorov, S. V. Fomin. Translated by Richard A. Silverman. Self-contained, evenly paced introduction to real and functional analysis. Some 350 problems. 403pp. 5⅜ x 8½.   0-486-61226-0

APPLIED ANALYSIS, Cornelius Lanczos. Classic work on analysis and design of finite processes for approximating solution of analytical problems. Algebraic equations, matrices, harmonic analysis, quadrature methods, much more. 559pp. 5⅜ x 8½.   0-486-65656-X

AN INTRODUCTION TO ALGEBRAIC STRUCTURES, Joseph Landin. Superb self-contained text covers "abstract algebra": sets and numbers, theory of groups, theory of rings, much more. Numerous well-chosen examples, exercises. 247pp. 5⅜ x 8½.   0-486-65940-2

QUALITATIVE THEORY OF DIFFERENTIAL EQUATIONS, V. V. Nemytskii and V.V. Stepanov. Classic graduate-level text by two prominent Soviet mathematicians covers classical differential equations as well as topological dynamics and ergodic theory. Bibliographies. 523pp. 5⅜ x 8½.   0-486-65954-2

THEORY OF MATRICES, Sam Perlis. Outstanding text covering rank, nonsingularity and inverses in connection with the development of canonical matrices under the relation of equivalence, and without the intervention of determinants. Includes exercises. 237pp. 5⅜ x 8½.   0-486-66810-X

INTRODUCTION TO ANALYSIS, Maxwell Rosenlicht. Unusually clear, accessible coverage of set theory, real number system, metric spaces, continuous functions, Riemann integration, multiple integrals, more. Wide range of problems. Undergraduate level. Bibliography. 254pp. 5⅜ x 8½.   0-486-65038-3

MODERN NONLINEAR EQUATIONS, Thomas L. Saaty. Emphasizes practical solution of problems; covers seven types of equations. ". . . a welcome contribution to the existing literature...."–*Math Reviews*. 490pp. 5⅜ x 8½.   0-486-64232-1

MATRICES AND LINEAR ALGEBRA, Hans Schneider and George Phillip Barker. Basic textbook covers theory of matrices and its applications to systems of linear equations and related topics such as determinants, eigenvalues and differential equations. Numerous exercises. 432pp. 5⅜ x 8½.   0-486-66014-1

LINEAR ALGEBRA, Georgi E. Shilov. Determinants, linear spaces, matrix algebras, similar topics. For advanced undergraduates, graduates. Silverman translation. 387pp. 5⅜ x 8½.   0-486-63518-X

ELEMENTS OF REAL ANALYSIS, David A. Sprecher. Classic text covers fundamental concepts, real number system, point sets, functions of a real variable, Fourier series, much more. Over 500 exercises. 352pp. 5⅜ x 8½.   0-486-65385-4

SET THEORY AND LOGIC, Robert R. Stoll. Lucid introduction to unified theory of mathematical concepts. Set theory and logic seen as tools for conceptual understanding of real number system. 496pp. 5⅜ x 8¼.   0-486-63829-4

CALCULUS OF VARIATIONS WITH APPLICATIONS, George M. Ewing. Applications-oriented introduction to variational theory develops insight and promotes understanding of specialized books, research papers. Suitable for advanced undergraduate/graduate students as primary, supplementary text. 352pp. 5⅜ x 8½.
0-486-64856-7

COMPLEX VARIABLES, Francis J. Flanigan. Unusual approach, delaying complex algebra till harmonic functions have been analyzed from real variable viewpoint. Includes problems with answers. 364pp. 5⅜ x 8½. 0-486-61388-7

AN INTRODUCTION TO THE CALCULUS OF VARIATIONS, Charles Fox. Graduate-level text covers variations of an integral, isoperimetrical problems, least action, special relativity, approximations, more. References. 279pp. 5⅜ x 8½.
0-486-65499-0

COUNTEREXAMPLES IN ANALYSIS, Bernard R. Gelbaum and John M. H. Olmsted. These counterexamples deal mostly with the part of analysis known as "real variables." The first half covers the real number system, and the second half encompasses higher dimensions. 1962 edition. xxiv+198pp. 5⅜ x 8½. 0-486-42875-3

CATASTROPHE THEORY FOR SCIENTISTS AND ENGINEERS, Robert Gilmore. Advanced-level treatment describes mathematics of theory grounded in the work of Poincaré, R. Thom, other mathematicians. Also important applications to problems in mathematics, physics, chemistry and engineering. 1981 edition. References. 28 tables. 397 black-and-white illustrations. xvii + 666pp. 6⅛ x 9¼.
0-486-67539-4

INTRODUCTION TO DIFFERENCE EQUATIONS, Samuel Goldberg. Exceptionally clear exposition of important discipline with applications to sociology, psychology, economics. Many illustrative examples; over 250 problems. 260pp. 5⅜ x 8½.
0-486-65084-7

NUMERICAL METHODS FOR SCIENTISTS AND ENGINEERS, Richard Hamming. Classic text stresses frequency approach in coverage of algorithms, polynomial approximation, Fourier approximation, exponential approximation, other topics. Revised and enlarged 2nd edition. 721pp. 5⅜ x 8½. 0-486-65241-6

INTRODUCTION TO NUMERICAL ANALYSIS (2nd Edition), F. B. Hildebrand. Classic, fundamental treatment covers computation, approximation, interpolation, numerical differentiation and integration, other topics. 150 new problems. 669pp. 5⅜ x 8½. 0-486-65363-3

THREE PEARLS OF NUMBER THEORY, A. Y. Khinchin. Three compelling puzzles require proof of a basic law governing the world of numbers. Challenges concern van der Waerden's theorem, the Landau-Schnirelmann hypothesis and Mann's theorem, and a solution to Waring's problem. Solutions included. 64pp. 5⅜ x 8½.
0-486-40026-3

THE PHILOSOPHY OF MATHEMATICS: AN INTRODUCTORY ESSAY, Stephan Körner. Surveys the views of Plato, Aristotle, Leibniz & Kant concerning propositions and theories of applied and pure mathematics. Introduction. Two appendices. Index. 198pp. 5⅜ x 8½. 0-486-25048-2

TENSOR CALCULUS, J.L. Synge and A. Schild. Widely used introductory text covers spaces and tensors, basic operations in Riemannian space, non-Riemannian spaces, etc. 324pp. 5⅜ x 8¼. 0-486-63612-7

ORDINARY DIFFERENTIAL EQUATIONS, Morris Tenenbaum and Harry Pollard. Exhaustive survey of ordinary differential equations for undergraduates in mathematics, engineering, science. Thorough analysis of theorems. Diagrams. Bibliography. Index. 818pp. 5⅜ x 8½. 0-486-64940-7

INTEGRAL EQUATIONS, F. G. Tricomi. Authoritative, well-written treatment of extremely useful mathematical tool with wide applications. Volterra Equations, Fredholm Equations, much more. Advanced undergraduate to graduate level. Exercises. Bibliography. 238pp. 5⅜ x 8½. 0-486-64828-1

FOURIER SERIES, Georgi P. Tolstov. Translated by Richard A. Silverman. A valuable addition to the literature on the subject, moving clearly from subject to subject and theorem to theorem. 107 problems, answers. 336pp. 5⅜ x 8½. 0-486-63317-9

INTRODUCTION TO MATHEMATICAL THINKING, Friedrich Waismann. Examinations of arithmetic, geometry, and theory of integers; rational and natural numbers; complete induction; limit and point of accumulation; remarkable curves; complex and hypercomplex numbers, more. 1959 ed. 27 figures. xii+260pp. 5⅜ x 8½. 0-486-63317-9

POPULAR LECTURES ON MATHEMATICAL LOGIC, Hao Wang. Noted logician's lucid treatment of historical developments, set theory, model theory, recursion theory and constructivism, proof theory, more. 3 appendixes. Bibliography. 1981 edition. ix + 283pp. 5⅜ x 8½. 0-486-67632-3

CALCULUS OF VARIATIONS, Robert Weinstock. Basic introduction covering isoperimetric problems, theory of elasticity, quantum mechanics, electrostatics, etc. Exercises throughout. 326pp. 5⅜ x 8½. 0-486-63069-2

THE CONTINUUM: A CRITICAL EXAMINATION OF THE FOUNDATION OF ANALYSIS, Hermann Weyl. Classic of 20th-century foundational research deals with the conceptual problem posed by the continuum. 156pp. 5⅜ x 8½. 0-486-67982-9

CHALLENGING MATHEMATICAL PROBLEMS WITH ELEMENTARY SOLUTIONS, A. M. Yaglom and I. M. Yaglom. Over 170 challenging problems on probability theory, combinatorial analysis, points and lines, topology, convex polygons, many other topics. Solutions. Total of 445pp. 5⅜ x 8½. Two-vol. set.
Vol. I: 0-486-65536-9 Vol. II: 0-486-65537-7

INTRODUCTION TO PARTIAL DIFFERENTIAL EQUATIONS WITH APPLICATIONS, E. C. Zachmanoglou and Dale W. Thoe. Essentials of partial differential equations applied to common problems in engineering and the physical sciences. Problems and answers. 416pp. 5⅜ x 8½. 0-486-65251-3

THE THEORY OF GROUPS, Hans J. Zassenhaus. Well-written graduate-level text acquaints reader with group-theoretic methods and demonstrates their usefulness in mathematics. Axioms, the calculus of complexes, homomorphic mapping, *p*-group theory, more. 276pp. 5⅜ x 8½. 0-486-40922-8

Physics

OPTICAL RESONANCE AND TWO-LEVEL ATOMS, L. Allen and J. H. Eberly. Clear, comprehensive introduction to basic principles behind all quantum optical resonance phenomena. 53 illustrations. Preface. Index. 256pp. 5⅜ x 8½. 0-486-65533-4

QUANTUM THEORY, David Bohm. This advanced undergraduate-level text presents the quantum theory in terms of qualitative and imaginative concepts, followed by specific applications worked out in mathematical detail. Preface. Index. 655pp. 5⅜ x 8½. 0-486-65969-0

ATOMIC PHYSICS (8th EDITION), Max Born. Nobel laureate's lucid treatment of kinetic theory of gases, elementary particles, nuclear atom, wave-corpuscles, atomic structure and spectral lines, much more. Over 40 appendices, bibliography. 495pp. 5⅜ x 8½. 0-486-65984-4

A SOPHISTICATE'S PRIMER OF RELATIVITY, P. W. Bridgman. Geared toward readers already acquainted with special relativity, this book transcends the view of theory as a working tool to answer natural questions: What is a frame of reference? What is a "law of nature"? What is the role of the "observer"? Extensive treatment, written in terms accessible to those without a scientific background. 1983 ed. xlviii+172pp. 5⅜ x 8½. 0-486-42549-5

AN INTRODUCTION TO HAMILTONIAN OPTICS, H. A. Buchdahl. Detailed account of the Hamiltonian treatment of aberration theory in geometrical optics. Many classes of optical systems defined in terms of the symmetries they possess. Problems with detailed solutions. 1970 edition. xv + 360pp. 5⅜ x 8½. 0-486-67597-1

PRIMER OF QUANTUM MECHANICS, Marvin Chester. Introductory text examines the classical quantum bead on a track: its state and representations; operator eigenvalues; harmonic oscillator and bound bead in a symmetric force field; and bead in a spherical shell. Other topics include spin, matrices, and the structure of quantum mechanics; the simplest atom; indistinguishable particles; and stationary-state perturbation theory. 1992 ed. xiv+314pp. 6⅛ x 9¼. 0-486-42878-8

LECTURES ON QUANTUM MECHANICS, Paul A. M. Dirac. Four concise, brilliant lectures on mathematical methods in quantum mechanics from Nobel Prize-winning quantum pioneer build on idea of visualizing quantum theory through the use of classical mechanics. 96pp. 5⅜ x 8½. 0-486-41713-1

THIRTY YEARS THAT SHOOK PHYSICS: THE STORY OF QUANTUM THEORY, George Gamow. Lucid, accessible introduction to influential theory of energy and matter. Careful explanations of Dirac's anti-particles, Bohr's model of the atom, much more. 12 plates. Numerous drawings. 240pp. 5⅜ x 8½. 0-486-24895-X

ELECTRONIC STRUCTURE AND THE PROPERTIES OF SOLIDS: THE PHYSICS OF THE CHEMICAL BOND, Walter A. Harrison. Innovative text offers basic understanding of the electronic structure of covalent and ionic solids, simple metals, transition metals and their compounds. Problems. 1980 edition. 582pp. 6⅛ x 9¼. 0-486-66021-4

CATALOG OF DOVER BOOKS

TENSOR CALCULUS, J.L. Synge and A. Schild. Widely used introductory text covers spaces and tensors, basic operations in Riemannian space, non-Riemannian spaces, etc. 324pp. 5⅜ x 8¼.   0-486-63612-7

ORDINARY DIFFERENTIAL EQUATIONS, Morris Tenenbaum and Harry Pollard. Exhaustive survey of ordinary differential equations for undergraduates in mathematics, engineering, science. Thorough analysis of theorems. Diagrams. Bibliography. Index. 818pp. 5⅜ x 8½.   0-486-64940-7

INTEGRAL EQUATIONS, F. G. Tricomi. Authoritative, well-written treatment of extremely useful mathematical tool with wide applications. Volterra Equations, Fredholm Equations, much more. Advanced undergraduate to graduate level. Exercises. Bibliography. 238pp. 5⅜ x 8½.   0-486-64828-1

FOURIER SERIES, Georgi P. Tolstov. Translated by Richard A. Silverman. A valuable addition to the literature on the subject, moving clearly from subject to subject and theorem to theorem. 107 problems, answers. 336pp. 5⅜ x 8½.   0-486-63317-9

INTRODUCTION TO MATHEMATICAL THINKING, Friedrich Waismann. Examinations of arithmetic, geometry, and theory of integers; rational and natural numbers; complete induction; limit and point of accumulation; remarkable curves; complex and hypercomplex numbers, more. 1959 ed. 27 figures. xii+260pp. 5⅜ x 8½.
0-486-63317-9

POPULAR LECTURES ON MATHEMATICAL LOGIC, Hao Wang. Noted logician's lucid treatment of historical developments, set theory, model theory, recursion theory and constructivism, proof theory, more. 3 appendixes. Bibliography. 1981 edition. ix + 283pp. 5⅜ x 8½.   0-486-67632-3

CALCULUS OF VARIATIONS, Robert Weinstock. Basic introduction covering isoperimetric problems, theory of elasticity, quantum mechanics, electrostatics, etc. Exercises throughout. 326pp. 5⅜ x 8½.   0-486-63069-2

THE CONTINUUM: A CRITICAL EXAMINATION OF THE FOUNDATION OF ANALYSIS, Hermann Weyl. Classic of 20th-century foundational research deals with the conceptual problem posed by the continuum. 156pp. 5⅜ x 8½.
0-486-67982-9

CHALLENGING MATHEMATICAL PROBLEMS WITH ELEMENTARY SOLUTIONS, A. M. Yaglom and I. M. Yaglom. Over 170 challenging problems on probability theory, combinatorial analysis, points and lines, topology, convex polygons, many other topics. Solutions. Total of 445pp. 5⅜ x 8½. Two-vol. set.
Vol. I: 0-486-65536-9   Vol. II: 0-486-65537-7